Metals as Clean Fuels

Acta Materialia Book Series

In collaboration with Acta Materialia Inc., this series complements Acta's portfolio of leading materials science journals published by Elsevier, delivering a full suite of content from foundational knowledge to primary research.

The **Acta Materialia Book Series** aims at providing a comprehensive, practical coverage of both core and advanced research topics in materials science. In addition to monographs on materials science fundamentals and new subfields in materials science, the book series features volumes on key practical aspects of research such as methods and protocols, supporting interdisciplinary research, and enabling the transition of new materials research into the applications realm.

Authored by leading experts, each monograph volume in the series will focus on a particular sub-field within materials science and engineering, frequently exemplified with specific case-studies, from structural materials, to advances in use-inspired device oriented functional materials and bio-inspired materials research.

ACTA MATERIALIA BOOK SERIES

Metals as Clean Fuels

ERIC DETSI
University of Pennsylvania, Philadelphia, PA, United States

JEFF TH. M. DEHOSSON
University of Groningen, Groningen, The Netherlands

Published in collaboration with Acta Materialia Inc.

Acta Materialia, Inc.
Publishing Leading Materials Science Journals Since 1953

ELSEVIER

Elsevier
Radarweg 29, PO Box 211, 1000 AE Amsterdam, Netherlands
125 London Wall, London EC2Y 5AS, United Kingdom
50 Hampshire Street, 5th Floor, Cambridge, MA 02139, United States

ISBN: 978-0-443-13537-8

For information on all Elsevier publications visit our website at https://www.elsevier.com/books-and-journals

Publisher: Matthew Deans
Acquisitions Editor: Ana Claudia Garcia
Editorial Project Manager: Isabella Silva
Production Project Manager: Maria Bernard
Cover Designer: Mark Rogers

Typeset by TNQ Tech

Contents

About the authors

Eric Detsi is currently an Associate Professor in Materials Science and Engineering at the University of Pennsylvania. He received his BSc (2006), MSc (2008), and PhD (with honors and highest distinction, 2012) degrees in Applied Physics at the University of Groningen in the Netherlands; afterward, he carried out his postdoctoral research in the Department of Chemistry at the University of California in Los Angeles (2013—16) as a Netherlands Science Foundation Rubicon Fellow.

He joined the University of Pennsylvania in 2016 as an Assistant Professor in Materials Science and Engineering. His current research involves electrochemical energy conversion and storage using nonconventional materials, such as liquid metals, nonprecious nanoporous metals, and disordered three-dimensional metal scaffolds. For details, see: https://directory.seas.upenn.edu/eric-detsi/.

Jeff Th. M. DeHosson received a PhD in Physics from the University of Groningen, the Netherlands (with honors and highest distinction), and after his postdoctoral years in the United States' (Northwestern University and UC Berkeley), he was appointed Professor in 1977 by the Crown (H.M. Queen Juliana). His passion is to carry out innovative and (pre-) competitive research in the field of materials sciences, with a particular emphasis on advances in (in situ) electron microscopy, lasers, and nanostructured materials, e.g., highly nanoporous materials, as started with Eric Detsi in 2008.

He is the Elected Member of the Royal Netherlands Academy of Sciences (division physics), the Royal Holland Society for Sciences and Humanities, and Academia Europaea; the Editor of *Acta Materialia* journals; and a Fellow of various scientific societies, including TMS, the United States' and American Society for Metals, FASM, the United States'. He acts as the Honorary Professor of Tsinghua University, Beijing; UST, Beijing; and the University of Port Elizabeth, SA. For his scientific research, he was honored with several international awards, including the European Materials Gold Medal, and he became Knighted by the Crown in 2019 (Knighthood, Netherlands Lion). For details, see https://www.rug.nl/staff/j.t.m.de.hosson/cv.

Preface

Climate change caused by greenhouse gas—emitting fossil fuels, alongside the recent surges in gas prices, has revived the urgency to reduce our heavy dependence on fossil fuels and to diversify our energy sources. Although various battery technologies, including Li-ion and Na-ion batteries, have emerged as promising alternatives to fossil fuels in recent years, unfortunately, practical batteries exhibit a relatively low energy density compared to fossil fuels. To illustrate this important issue in a bit more detail by the present state of the art, the current energy density of Li-ion batteries using graphite as the anode and lithium iron phosphate as the cathode is around 0.16 kWh/kg, which is approximately two orders of magnitude lower than the energy density of fossil fuels! An attractive route to address this crucial drawback of existing batteries is to go for metals as clean fuels. Not only are metal fuels more energy-dense than fossil fuels, but importantly they can generate energy without CO_2 emissions, unlike fossil fuels. This monograph aims at paving the way to a paradigm shift toward metals as clean fuels for specific applications.

Indeed, metals and metalloids that can spontaneously react with oxygen or water to produce energy without releasing carbon dioxide have been proposed as clean fuels for their high volumetric energy densities in recent years. For example, several proof of concepts demonstrating the use of aluminum as a fuel for power generation using water as the oxidant have been reported, including an experimental aluminum-fueled power plant, and kilowatt-scale fuel cell systems powered by recycled aluminum. Despite these promising concepts, the emergence of the metal fuel economy has been impeded by a major bottleneck involving the activation of metal fuels in a scalable manner to render them highly reactive with water and safe to use. Although thermodynamically, metal fuels can spontaneously react with O_2 and H_2O to release energy, in practice, these reactions are kinetically extremely sluggish and, therefore, will not really have a great impact on practical service.

This monograph presents several advanced novel ideas about removing these principal obstacles by focusing on activating metal fuels through nanostructuring, which is aimed at increasing the surface-to-volume ratio of the metal to enhance the density of active metal sites. As an exciting example: in particular, the content of this monograph will show that the only way to execute hydrolysis is by using nanoscale material with ultrasmall size in the range of 5–15 nm.

To structure the content of this book, the authors decided to discuss metals that are on different ends of the energy density scale, e.g., aluminum has a very high energy density but converting the reaction product (aluminum oxide) back to aluminum metal in a sustainable way is still a challenge. In contrast, zinc has a relatively low energy density, but the reaction product (Zn oxide) can easily be converted back to Zn metal in a sustainable manner. Therefore, those two metals at both ends of the "energy density" scale will be discussed so as to inspire the community.

Special attention is also paid to the scalability and value-added byproduct of hydrogen gas production via the hydrolysis reactions. A 20-fold improvement in the dealloying rate to produce nanoporous aluminum is demonstrated in the monograph using various strategies: increasing the ionic conductivity of the electrolyte, the surface area of the electrodes, et cetera.

In the monograph, besides scalability, reduction of CO_2 emission, and aspects of energy density as such, attention is also paid to novel sustainable methods for the fabrication of metal/metal nanocomposites without losing sacrificial materials. This sustainable synthesis approach not only reduces the loss of material during dealloying but also opens new avenues for developing metal/metal nanocomposites made of dissimilar nanoporous metal layers stacked together. It offers significant implications for applications in catalysis and beyond, in which the synergistic effect between the underlying nanoporous metal layers could enhance the properties of the nanocomposite. Finally, this book shows that, besides chemical composition and geometrical defects like interfaces, topology is an extra design parameter, in addition to extrinsic and intrinsic size-dependent material properties. An inspiring methodology based on *Integral-Geometry Morphological Image Analysis* of porous material systems is presented and discussed. It is demonstrated that the method serves as an attractive toolset for analyzing geometrical objects and topological patterns in various porous systems.

The monograph is targeted at senior undergraduate, PhD research students, postgraduate researchers, and anyone interested in entering this field of nanostructured materials.

As such, the monograph would be the first one on the market that presents state-of-the-art nanostructured metals as clean fuels, and likewise, this timely book could catalyze and advance research in this field, which sits at the center of a myriad materials science disciplines coming together to form highly versatile material systems for energy applications.

Eric Detsi (UPENN, USA) and Jeff Th. M. DeHosson (Groningen, the Netherlands)

Acknowledgments

We are grateful to our team of highly talented PhD students, master's degree students, undergraduate students, and postdoctoral researchers for their enlightening discussions, sparkling contributions, creativity, ingenuity, and hard work over the past years.

The authors are deeply indebted to Jintao Fu (currently a Postdoctoral Research Associate at the Materials Synthesis and Integrated Devices (MPA-11) Group at Los Alamos National Laboratory) for proofreading and his enthusiastic and professional help in completing this book together with postdoctoral fellow Mohamed Hassan at UPENN. AI-assisted proofreading was used to improve some sections of this book using *Grammarly.*

As an introductory text to the state-of-the-art field of research reference is made to our reviews published by Jintao Fu, Sam Welborn, and Eric Detsi in *ACS Applied Energy Materials Vol. 5* (2022) Invited Contribution, entitled *Dealloyed Air- and Water-Sensitive Nanoporous Metals and Metalloids for Emerging Energy Applications,* and by Jintao Fu, Eric Detsi, Jeff Th.M. De Hosson, entitled *Recent advances in nanoporous materials for renewable energy resources conversion into fuels*, in *Surface and Coatings Technology Vol. 347* (2018) 320–336 (invited).

The authors particularly thank Jintao Fu, John Corsi, Sam Welborn, and Alexander Ng for their discussion and contributions to Chapter 3 (see above) as well as John S. Corsi's dissertation under the supervision of E. Detsi as input to this chapter, entitled *Fabrication and Study of Microstructure, Solid-Electrolyte Interphase, And Phase Evolution of Nanoporous Metals Used as Li-Ion Battery Alloy Anodes* (UPenn, 2021).

https://repository.upenn.edu/entities/publication/39e8d876-f749-4140-916b-6728148bcf63

The authors are grateful to Jintao Fu, Ziling Deng, Timothy Lee, John Corsi, Zeyu Wang, and Dongyang Zhang for their discussion and contributions presented to Chapter 4. In particular, part of the content of Chapter 4 has been published by Jintao Fu, Ziling Deng, Timothy Lee, John Corsi, Zeyu Wang, Dongyand Zhang, and Eric Detsi in ACS *Applied Energy Materials Vol. 1* (2018), entitled *pH-Controlled Dealloying Route to Hierarchical Bulk Nanoporous Zn Derived from Metastable Alloy for Hydrogen Generation by Hydrolysis of Zn in Neutral Water.*

The authors are very thankful to Jo Kubota and Jiaxin Liu for their contributions to Chapter 5. Part of the content of Chapter 5 has been submitted by Jo Kubota, Jiaxin Liu, and Eric Detsi for publication in ACS *Applied Materials & Interfaces* (2025), entitled, *Rapid Synthesis of Nanoporous Zn Powder by Selective Etching of Al from Micrometer-Sized Zn-Al Powder Particles Produced by Gas-Atomization and Its Application in Hydrogen Generation.*

The authors are grateful to John Corsi, Jintao Fu, Zeyu Wang, Timothy Lee, and Alexander Ng for their discussion and contributions to Chapter 6. In particular, part of the content of Chapter 6 has been published in ACS *Sustainable Chemistry & Engineering Vol.* 7 (2019), entitled, *Hierarchical Bulk Nanoporous Aluminum for On-Site Generation of Hydrogen by Hydrolysis in Pure Water and Combustion of Solid Fuels.*

The authors are thankful to Timothy Lee, Jintao Fu, Victoria Basile, John S. Corsi, Zeyu Wang, Hyeongjun Koh, Alexander K. Ng, Jiaxin Liu, and Eric A. Stach for their discussion and contributions to the work presented in Chapter 7. Part of the content of Chapter 7 has been published by Timothy Lee, Jintao Fu, Victoria Basile, John S. Corsi, Zeyu Wang, and Eric Detsi in *Renewable Energy Vol. 155* (2020), entitled *Activated alumina as value-added byproduct from the hydrolysis of hierarchical nanoporous aluminum with pure water to generate hydrogen fuel,* and by Timothy Lee, Hyeongjun Koh, Alexander K. Ng, Jiaxin Liu, Eric A. Stach, and Eric Detsi, in *Scripta Materialia Vol. 221* (2022), entitled *Ultrafine nanoporous aluminum by electrolytic dealloying of aluminum-magnesium alloys in glyme-based electrolytes with recovery of sacrificial magnesium.*

The authors are indebted to Alexander K. Ng and Hyeongjun Koh for their discussion and contributions to Chapter 8. In particular, the content of Chapter 8 has been submitted for publication by Alexander K. Ng, Hyeongjun Koh, and Eric Detsi, in *Acta Materialia*, entitled *Tri-layer nanoporous silver | silver | silver by etching without sacrificing materials through the Kirkendall effect.* The authors are thankful to Alexander K. Ng and Hyeongjun Koh for their discussion and contributions presented in Chapter 9. The content of Chapter 9 has been submitted for publication by Alexander K. Ng, Hyeongjun Koh, and Eric Detsi in *Acta Materialia*, entitled *Tri-layer nanoporous silver | gold | silver by etching without sacrificing materials through the Kirkendall effect.*

The authors are thankful to Kristel Michielsen and Hans DeRaedt for fruitful collaborations and for their interest in porous media. Part of the content of Chapter 10 is based upon joint work by Kristel F.L. Michielsen, Hans DeRaedt, and Jeff Th.M. DeHosson, published in *Advances in Imaging and Electron Physics Volume 125* (2002), entitled *Aspects of Mathematical Topology.*

The authors acknowledge the financial support and facilitation during the research of porous materials systems from various funding organizations over the past decade. In particular, the authors would like to express their gratitude to the U.S. National Science Foundation (NSF), Division of Civil Mechanical and Manufacturing Innovation (CMMI) EArly-concept Grant for Exploratory Research (EAGER) award #1840672, NSF Division of Materials Research (DMR) Future Manufacturing award #2134715, and NSF-CMMI Faculty Early Career Development Program (CAREER) award #2047851. They also acknowledge NSF for funding the Dual Source (Cu and Mo) and Environmental X-ray Scattering (DEXS) facility under NSF MRI #1725969 used to generate X-ray scattering data presented in this book.

Additionally, some data presented in this book were collected in part at the Singh Center for Nanotechnology, which is supported by the NSF grant NNCI #2025608 as part of the National Nanotechnology Coordinated Infrastructure (NNCI) Program.

The authors are grateful to the Netherlands Organization for Scientific Research (NWO-the Hague), Materials Innovation Institute (M2i), and the Zernike Institute of Advanced Materials (ZiAM, Groningen) for their financial support; the work was partly carried out within the framework of the Strategic Research Program of the Netherlands Institute for Metals Research and the Foundation for Fundamental Research on Matter, FOM-Utrecht (# 02EMM22).

CHAPTER ONE

How metal fuels work

1.1 Overview

Climate change caused by the emission of greenhouse gases from fossil fuels such as petroleum, natural gas, and coal, along with the recent surges in gas prices due to geopolitical tensions, underscores the need to reduce our heavy dependence on fossil fuels and diversify our energy supply chains. Although various battery technologies, including Li-ion and Na-ion batteries, have emerged as promising alternatives to fossil fuels, unfortunately, practical Li- and Na-ion batteries exhibit a relatively low energy density compared to fossil fuels (Fig. 1.1) [1,2]. For example, the current energy density of state-of-the-art Li-ion batteries using graphite as the anode and lithium iron phosphate ($LiFePO_4$ or LFP) as the cathode is around 0.16 kWh/kg [3], which is approximately two orders of magnitude lower

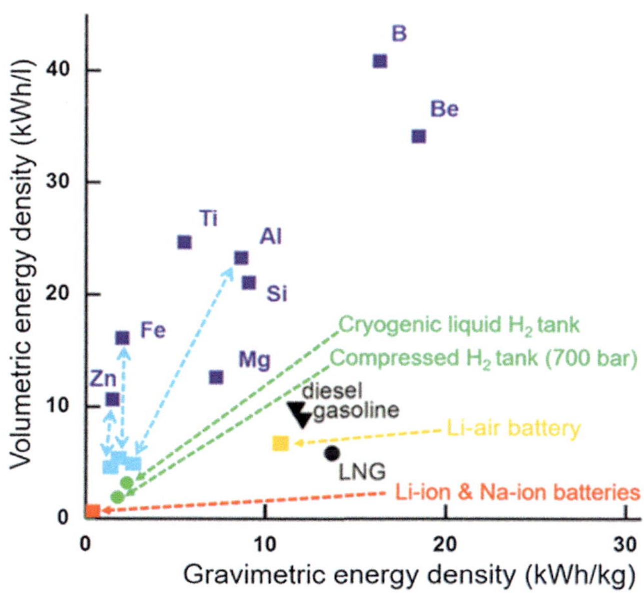

Figure 1.1 Comparison of the energy density by weight versus volume of fossil fuels (black), metal fuels (blue), hydrogen fuel (green), and Li-ion battery (red). *Adapted from Refs. [1,2], with permission.*

Metals as Clean Fuels
ISBN: 978-0-443-13537-8
https://doi.org/10.1016/B978-0-443-13537-8.00008-7

than the energy density of fossil fuels. Obviously meeting the global energy demand will require diversifying our energy supply chains by embracing nonconventional energy storage systems that operate under nonstandard conditions, such as at high temperatures and pressures, and that are cleaner, more energy-dense, and more scalable than conventional energy storage technologies. One such nonconventional energy storage system that will be the focus of this monograph is *metals as clean fuels*. Not only are metal fuels more energy-dense than fossil fuels, but importantly, they can generate energy without CO_2 emissions, unlike fossil fuels. This monograph aims to pave the way to a paradigm shift toward metals as a clean fuel for specific applications.

1.2 Why use metals as clean fuels?

Unlike CO_2-emitting fossil fuels, several metals and metalloids can spontaneously react with O_2 (i.e., metal combustion or dry metal fuel oxidation) and H_2O (i.e., metal hydrolysis or wet metal fuel oxidation) to release energy without CO_2 emissions [4,5]. In Fig. 1.1, adapted from Refs. [1,2], we compare the gravimetric and volumetric energy densities of various fuel systems, including fossil fuels (black); dry metal fuels (dark blue) calculated based on the mass and volume of the metals; wet metal fuels (light blue) including Al, Fe, Zn, and calculated based on the stoichiometric mass and volume of the metal-water mixture; compressed and liquefied hydrogen fuels (green) calculated based on the mass of the hydrogen and its storage tank; Li-air battery (orange); and finally Li- and Na-ion batteries (red) [1,2]. It is seen that the volumetric energy density of most dry metal fuels is significantly higher than that of fossil fuels, hydrogen fuels, and Li- and Na-ion batteries. Additionally, except for Fe and Zn, the gravimetric energy density of most dry metal fuels is comparable to that of fossil fuels and significantly higher than that of compressed and liquefied hydrogen fuel systems and that of Li- and Na-ion batteries.

Note that although Li-air batteries (orange) have the potential to compete with fossil fuels and some metal fuels in terms of gravimetric and volumetric energy densities, practical Li-air batteries have yet to be developed. Aside from dry metal fuels (dark blue), wet metal fuels (light blue) significantly outperform Li- and Na-ion batteries in terms of gravimetric and volumetric energy densities. Further, wet metal fuels have a higher volumetric energy density than compressed and liquefied hydrogen fuel systems. Wet and dry metal fuels can be used as complementary fuels: wet metal fuels operate at low-to-moderate temperatures and/or pressures, making them suitable for applications that only withstand mild conditions, while dry metal fuels operate at high temperatures and/or pressures, making them ideal for applications that can tolerate harsher conditions.

1.3 A key challenge with metal fuels

Despite their high energy density and potential to generate energy without CO_2 emissions, metal fuels face significant obstacles that hinder their use in practical applications. One of these obstacles, which is the focus of this monograph, is the *activation of metal fuels*. Although thermodynamically, metal fuels can spontaneously react with O_2 and H_2O to release energy, in practice, these reactions are kinetically sluggish and will only proceed if the metal fuels are activated.

There are three common methods to activate metal fuels, discussed in detail in Chapter 2: (i) *Chemical activation*, during which a catalyst is used to lower the activation energy barrier for the reaction, boosting in that way the reaction rate [6]. (ii) *Thermal and/or mechanical activation*, during which the temperature and/or pressure of the system is increased to enhance the reaction rate [7]. (iii) *Nanostructuring*, which is aimed at increasing the surface-to-volume ratio of the metal to enhance the density of active metal sites to facilitate the reaction between the metal and O_2 or H_2O [8–11]. This monograph will focus on the latter approach, activating metal fuels through nanostructuring, discussed in Chapters 3–10. In the following sections, we will explain how metal fuels work by comparing them with the working principle of batteries.

1.4 Working principle of batteries

The primary objective of the current chapter is to elucidate the working principle of metal fuels. In fact, we can better understand how metal fuels work by comparing them with Li and Na batteries. Both batteries and metal fuels store and release energy through chemical reactions involving metals. The fundamental principle that governs the working of a battery is the difference in electron energy levels between the metal anode (e.g., Li and Na metals), in which electrons occupy a higher energy level (Fermi energy level), and the nonmetal cathode (e.g., $LiCoO_2$, $LiMn_2O_4$, $LiFePO_4$ for Li-ion batteries, and $NaFePO_4$, $Na_{2/3}Ni_{1/3}Mn_{2/3}O_2$ for Na-ion batteries), in which electrons occupy a lower energy level. Since in a closed system, electrons tend to occupy a lower energy level, if the anode and cathode of a battery are immersed in a suitable electrolyte and electrically connected externally to close the loop, electrons will spontaneously move through the external electrical connections from the higher energy level in the anode to the lower energy level in the cathode. During this process, the cathode gaining electrons is reduced, while Li (or Na) metal losing electrons at the anode is oxidized to Li^+ (or Na^+). The Li^+ (or Na^+) is then transferred through the electrolyte to the cathode, where it is intercalated and recombined with the electron from the anode [12,13].

The reduction reaction at the cathode, known as the *discharge* (green, Fig. 1.2), can release two forms of energy: electrical energy, at most equivalent to the energy levels difference between the anode and cathode, and thermal energy (i.e., energy loss during discharge) that depends on the discharge rate. The discharge of electrical energy $\left(E_{\text{discharge}}\right)$ can be quantified using the relation in Eq. (1.1):

$$E_{\text{discharge}} \leq q(V_C - V_A) \tag{1.1}$$

where q represents the total charge transferred between the anode and the cathode, V_C and V_A are the electrical voltage at the cathode and anode, respectively, with $V_C > V_A$. From a thermodynamic viewpoint, the reduction reaction that governs the discharge of a battery is spontaneous with a negative change in free energy $\left(\Delta G_{\text{discharge}} < 0\right)$ equal to $-E_{\text{discharge}}$, assuming that there is no energy loss in the form of heat. Under these conditions, Eq. (1.1) can be rewritten as Eq. (1.2) [14]:

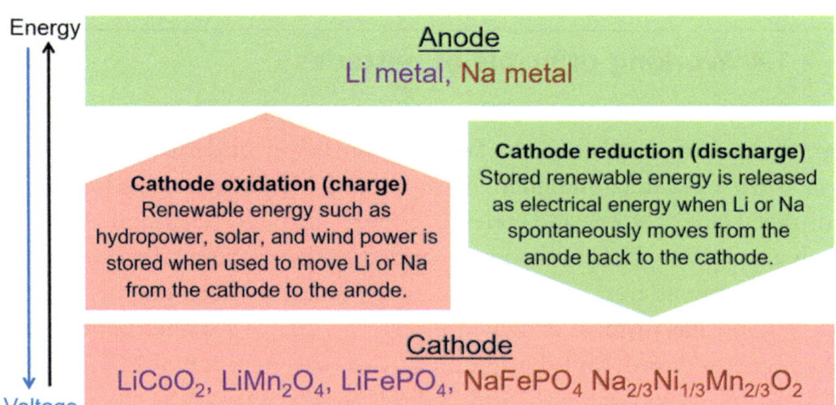

Figure 1.2 How Li-ion and Na-ion batteries work.

$$\Delta G_{\text{discharge}} = -q(V_C - V_A) \tag{1.2}$$

During the reverse process, moving electrons from a lower energy level at the cathode to a higher energy level at the anode is not a spontaneous process since electrons naturally go down in energy, not up. Therefore, external electrical energy is required to transfer electrons from the cathode to the anode. In this case, the cathode loses electrons and gets oxidized, which leads to the release of Li (or Na) from the cathode in the electrolyte in the form of Li^+ (or Na^+). The Li^+ (or Na^+) is transferred through the electrolyte to the anode, where it combines with the electron from the cathode external circuit to form deposits of Li (or Na) metal. This cathode oxidation reaction, known as the *charge* (pink, Fig. 1.2), requires an electrical energy supply that should be greater (to compensate for energy loss during charging) than the energy level difference between the anode and cathode. This electrical energy (E_{charge}) can be quantified using Eq. (1.3):

$$E_{\text{charge}} \geq q(V_C - V_A) \tag{1.3}$$

where q, V_C, and V_A are the same as defined earlier for the discharge step, with $V_C > V_A$. From a thermodynamic viewpoint, the oxidation reaction that governs the charge of a battery is nonspontaneous with a positive change in free energy $(\Delta G_{\text{charge}} > 0)$ equal to E_{charge}, assuming that there is no energy loss in the form of heat. Under these conditions, Eq. (1.3) can be rewritten as Eq. (1.4):

$$\Delta G_{\text{charge}} = q(V_C - V_A) \tag{1.4}$$

If the energy used to charge the battery is supplied from renewable sources such as hydropower, solar power, and wind power, then it can be argued that the battery stores renewable energy during charging. As a result, the electrical energy released during the discharge step is considered *green energy*.

1.5 Working principle of dry metal fuels

First, let us examine the high-energy density metal-air battery in Fig. 1.1 (orange), where Li metal is used at the anode, and oxygen (O_2) gas from air is used at the cathode. During the discharge in a nonaqueous electrolyte, the Li metal anode loses electrons and undergoes oxidation to become Li^+, which is transferred through the electrolyte to the cathode, where it combines with the electron from the anode external circuit and O_2 from air to form metal oxides such as Li_2O and Li_2O_2 and releases electrical energy, $E_{discharge}$ [15]. As discussed earlier, since the reaction that governs the discharge of a battery is spontaneous $(\Delta G_{discharge} < 0)$, the oxidation of Li metal in the presence of O_2 to form Li_2O and Li_2O_2 can naturally proceed, even without the need for a battery, as in the case of metal combustion. When metal oxidation/combustion naturally occurs without a battery, the energy is not released as electrical energy $(E_{discharge})$, but as heat, which we refer to as $Q_{thermal}$. Aside from Li metal used here for the sake of illustration, under suitable conditions, most reactive metals (e.g., Mg, Al, Fe, Zn, etc.) and metalloids (e.g., B, Si, etc.) can also spontaneously oxidize/combust in the presence of O_2, converting in that way chemical energy into thermal energy, like fossil fuels [4]. More importantly, these reactive metals and metalloids are safer and more practical for use as metal fuels than Li and other alkali metals. Therefore, our focus will now shift from Li to Al, Fe, and Zn as three examples of practical metal fuels. It is worth emphasizing that these metal fuels (Al, Fe, Zn) are cleaner than fossil fuels since their oxidation products are metal oxides (Al_2O_3, Fe_3O_4/Fe_2O_3, and ZnO) instead of carbon dioxide, as with fossil fuels. The free energy decreases $(\Delta G_{oxidation} < 0)$, when metallic Al, Fe, or Zn oxidizes/combusts in air and forms Al_2O_3, Fe_3O_4/Fe_2O_3, and ZnO, respectively (green, Fig. 1.3). The energy difference is released as heat, $Q_{thermal}$, through Eq. (1.5):

$$\Delta G_{oxidation} = -Q_{thermal} \tag{1.5}$$

The chemical reactions that take place during the combustion of Al, Fe, and Zn in the presence of O_2 are provided in Eqs. (1.6)–(1.8) in Table 1.1. This table also includes the corresponding gravimetric (8.4 kWh/kg,

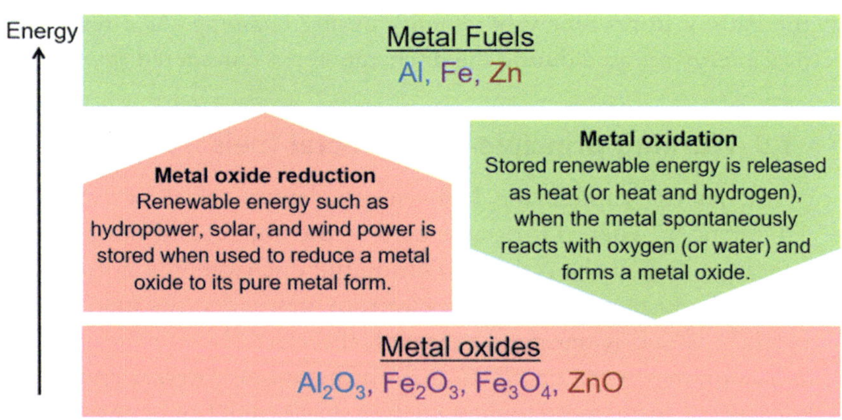

Figure 1.3 How metal fuels work.

Table 1.1 Overview of chemical reactions during the combustion of Al, Fe, and Zn in the presence of O_2 and heat release.

Dry metal fuel oxidation	$Q_{thermal}$	Eq. #
$4Al\ (s) + 3O_2\ (g) \rightarrow$ $2Al_2O_3\ (s) + Q_{thermal}$	$8.4\ \frac{kWh}{kg_{Al}}$ $22.7\ \frac{kWh}{l_{Al}}$	(1.6)
$4Fe\ (s) + 3O_2\ (g) \rightarrow$ $2Fe_2O_3\ (s) + Q_{thermal}$	$2.0\ \frac{kWh}{kg_{Fe}}$ $16.1\ \frac{kWh}{l_{Fe}}$	(1.7)
$2Zn\ (s) + O_2\ (g) \rightarrow$ $ZnO\ (s) + Q_{thermal}$	$1.4\ \frac{kWh}{kg_{Zn}}$ $9.6\ \frac{kWh}{l_{Zn}}$	(1.8)

2.0 kWh/kg, and 1.4 kWh/kg) and volumetric (22.7 kWh/l, 16.1 kWh/l, and 9.6 kWh/l) thermal energies, $Q_{thermal}$, released per kilogram and per liter l of Al, Fe, and Zn, respectively.

During the reverse reduction process (pink, Fig. 1.3), the metal oxides (Al_2O_3, Fe_3O_4/Fe_2O_3, ZnO) release oxygen to regenerate the corresponding metals (Al, Fe, Zn). When a metal oxide is reduced to its pure metal form, the reaction is nonspontaneous, since the metal oxide is thermodynamically more stable than its pure metal counterpart, resulting in a positive change in free energy ($\Delta G_{reduction} > 0$). Energy is therefore required to reduce a metal oxide to metal. This energy is usually supplied as thermal energy in conjunction with a reducing agent or as thermal energy in conjunction with electrical energy. There are two most popular methods (among many) to convert metal oxides to metals using heat and a reducing agent. The first method is carbothermic reduction, which often uses carbon or methane as the reducing agent [16]. In industry, Fe is commonly produced from Fe_3O_4/Fe_2O_3 using carbothermic reduction. However, this method releases CO_2, making it unsuitable for clean metal fuel applications. The second method is metallothermic reduction, such as the Kroll process [17] and the Armstrong process [18], which use a reactive metal as the reducing agent.

Similarly, the two most popular methods for converting metal oxides to metals using heat and electricity are molten oxide electrolysis, such as the Hall-Héroult process [19], and direct electrochemical reduction of metal oxides, such as the FFC-Cambridge process [20,21]. The former method reduces the metal oxide to metal in the liquid state. In industry, Al is produced from Al_2O_3 using the Hall-Héroult process, but because this method is highly carbon-intensive, it is not suitable for clean metal fuel applications. ELYSIS (a joint venture of two prominent Al industry pioneers,

Alcoa and Rio Tinto, supported by Apple, the Canadian Government, and the Quebec Government) has improved the Hall-Héroult process to reduce molten Al_2O_3 to Al without CO_2 emissions, so-called carbon-free Al smelting [22]. The latter method reduces the metal oxide to metal in the solid state using a molten salt electrolyte. It is important to note that electrolysis and electrochemical reduction of metal oxides do not typically occur in aqueous solutions due to the competing hydrogen evolution reaction, which obstructs metal deposition. However, certain metals such as Zn do not intrinsically catalyze hydrogen evolution, which allows for the deposition of Zn metal from an aqueous solution at room temperature in industry [23]. Therefore, Zn is particularly suitable as a recyclable clean metal fuel. As such, the use of Zn as a metal fuel will be extensively discussed in this monograph (see Chapters 2 and 4).

In summary for this section, under suitable conditions (these conditions include metal fuels activation, which will be discussed in Chapter 2), metal fuels can react with O_2 from air to release energy in the form of heat, transforming into metal oxides that can be converted back into metals for reuse. The process of converting metal oxides back to metals can be compared to charging a battery. If renewable energy is used to convert metal oxides to metals, and the process does not involve greenhouse gas emissions, it can be considered that metal fuels store renewable energy during this reduction process. As a result, the thermal energy released during dry metal fuel oxidation can be considered *green energy*.

1.6 Working principle of wet metal fuels

In addition to releasing energy from metal fuels through dry oxidation/combustion, during which O_2 gas is used as the oxidizer, metal fuels can also release energy through wet oxidation, referred to as metal hydrolysis as well, during which a liquid, often H_2O, is used as the oxidizer. While during dry metal fuel oxidation, all the chemical energy from the reaction is converted into thermal energy, during wet metal fuel oxidation, the chemical energy from the reaction is converted into thermal energy and a gaseous chemical fuel (hydrogen $-H_2$), which carries a new form of chemical energy [24]. In this section, we discuss the reaction of H_2O with Al, Fe, and Zn as practical examples of wet metal fuels, as we did with dry metal fuels in Section 1.5. In general, metal-H_2O reactions release heat, chemical fuel (i.e., H_2), and a solid by-product whose nature depends on the reaction temperature and/or pressure. For example, while $Fe-H_2O$ and $Zn-H_2O$

reactions produce Fe_3O_4 and ZnO, respectively, as solid by-products at near room temperatures and moderate temperatures, Al—H_2O reaction produces $Al(OH)_3$, AlOOH, and Al_2O_3 as solid by-products at near room temperatures, moderate temperatures, and high temperatures, respectively [25].

Like with dry metal fuels, the free energy decreases ($\Delta G_{oxidation} < 0$) when Al, Fe, and Fe react with H_2O to release heat, H_2, and form the following solid by-products $Al(OH)_3$, Fe_3O_4, and ZnO, respectively (Fig. 1.3). The energy difference is released as heat, $Q_{thermal}$ and chemical energy $Q_{chemical}$ (or chemical fuel) through Eq. (1.9):

$$\Delta G_{oxidation} = -(Q_{thermal} + Q_{chemical}) \tag{1.9}$$

The chemical reactions that take place during the hydrolysis of Al, Fe, and Zn are given in Eqs. 1.10—1.12 from Table 1.2, together with the corresponding values of the gravimetric and volumetric thermal energy $Q_{thermal}$ and chemical energy $Q_{chemical}$ released per kilogram and per liter of Al, Fe, and Zn.

To regenerate the metal (pink, Fig. 1.3), in some cases, the metal oxide solid by-products (e.g., Fe_3O_4/Fe_2O_3 and ZnO) can be directly reduced back to the metal form, as discussed in Section 1.5 above. In other cases, the metal hydroxide solid by-products (e.g., $Al(OH)_3$ and AlOOH) first need to be converted into metal oxides, prior to reducing them back to metal.

Table 1.2 Chemical reactions that take place during the hydrolysis of Al, Fe, and Zn, together with the corresponding values of the gravimetric and volumetric thermal energy $Q_{thermal}$ and chemical energy $Q_{chemical}$ released per kilogram and per liter of Al, Fe, and Zn.

Wet metal fuel oxidation	$Q_{thermal}$	$Q_{chemical}$	$Q_{thermal} + Q_{chemical}$	Eq. #
$2Al(s) + 6H_2O(l) \rightarrow 2Al(OH)_3(s) + 3H_2(g)$ (or $Q_{chemical}$) + $Q_{thermal}$	$1.4\ \frac{kWh}{kg_{Al} + H_2O}$ $1.8\ \frac{kWh}{l_{Al} + H_2O}$	$1.4\ \frac{kWh}{kg_{Al} + H_2O}$ $1.8\ \frac{kWh}{l_{Al} + H_2O}$	$2.8\ \frac{kWh}{kg_{Al} + H_2O}$ $3.6\ \frac{kWh}{l_{Al} + H_2O}$	(1.10)
$3Fe(s) + 4H_2O(l) \rightarrow Fe_3O_4(s) + 4H_2(g)$ (or $Q_{chemical}$) + $Q_{thermal}$	$0.1\ \frac{kWh}{kg_{Fe} + H_2O}$ $0.3\ \frac{kWh}{l_{Fe} + H_2O}$	$1.3\ \frac{kWh}{kg_{Fe} + H_2O}$ $3.5\ \frac{kWh}{l_{Fe} + H_2O}$	$1.4\ \frac{kWh}{kg_{Fe} + H_2O}$ $3.8\ \frac{kWh}{l_{Fe} + H_2O}$	(1.11)
$Zn(s) + H_2O(l) \rightarrow ZnO(s) + H_2(g)$ (or $Q_{chemical}$) + $Q_{thermal}$	$0.2\ \frac{kWh}{kg_{Zn} + H_2O}$ $0.4\ \frac{kWh}{l_{Zn} + H_2O}$	$0.9\ \frac{kWh}{kg_{Zn} + H_2O}$ $2.9\ \frac{kWh}{l_{Zn} + H_2O}$	$1.1\ \frac{kWh}{kg_{Zn} + H_2O}$ $3.3\ \frac{kWh}{l_{Zn} + H_2O}$	(1.12)

References

[1] P. Julien, J.M. Bergthorson, Enabling the metal fuel economy: green recycling of metal fuels, Sustain. Energy Fuels 1 (2017) 615–625, https://doi.org/10.1039/C7SE00004A.

[2] D.J. Durbin, C. Malardier-Jugroot, Review of hydrogen storage techniques for on board vehicle applications, Int. J. Hydrogen Energy 38 (2013) 14595–14617, https://doi.org/10.1016/j.ijhydene.2013.07.058.

[3] M.S. Whittingham, Ultimate limits to intercalation reactions for lithium batteries, Chem. Rev. 114 (2014) 11414–11443, https://doi.org/10.1021/cr5003003.

[4] J.M. Bergthorson, S. Goroshin, M.J. Soo, P. Julien, J. Palecka, D.L. Frost, D.J. Jarvis, Direct combustion of recyclable metal fuels for zero-carbon heat and power, Appl. Energy 160 (2015) 368–382, https://doi.org/10.1016/j.apenergy.2015.09.037.

[5] J.M. Bergthorson, Recyclable metal fuels for clean and compact zero-carbon power, Prog. Energy Combust. Sci. 68 (2018) 169–196, https://doi.org/10.1016/j.pecs.2018.05.001.

[6] A. V Ilyukhina, A.S. Ilyukhin, E.I. Shkolnikov, Hydrogen generation from water by means of activated aluminum, Int. J. Hydrogen Energy 37 (2012) 16382–16387, https://doi.org/10.1016/j.ijhydene.2012.02.175.

[7] Y. Kanda, S. Kondo, S. Ooya, T. Kobayashi, Y. Uemichi, K. Higashino, M. Sugioka, Green hydrogen production by mechanical mixing of aluminum with water, J. Chem. Eng. Jpn. 44 (2011) 803–808, https://doi.org/10.1252/jcej.10we272.

[8] J. Fu, S.S. Welborn, E. Detsi, Dealloyed air- and water-sensitive nanoporous metals and metalloids for emerging energy applications, ACS Appl. Energy Mater. 5 (2022) 6516–6544, https://doi.org/10.1021/acsaem.2c00405.

[9] J.S. Corsi, J. Fu, Z. Wang, T. Lee, A.K. Ng, E. Detsi, Hierarchical bulk nanoporous aluminum for on-site generation of hydrogen by hydrolysis in pure water and combustion of solid fuels, ACS Sustain. Chem. Eng. 7 (2019) 11194–11204, https://doi.org/10.1021/acssuschemeng.9b00481.

[10] J. Fu, Z. Deng, T. Lee, J.S. Corsi, Z. Wang, D. Zhang, E. Detsi, pH-controlled dealloying route to hierarchical bulk nanoporous Zn derived from metastable alloy for hydrogen generation by hydrolysis of Zn in neutral water, ACS Appl. Energy Mater. 1 (2018) 3198–3205, https://doi.org/10.1021/acsaem.8b00419.

[11] J. Fu, J.S. Corsi, S.S. Welborn, V. Basile, L. Wang, A.K. Ng, E. Detsi, Eco-friendly synthesis of nanoporous magnesium by air-free electrolytic dealloying with recovery of sacrificial elements for energy conversion and storage applications, ACS Sustain. Chem. Eng. 9 (2021) 2762–2769, https://doi.org/10.1021/acssuschemeng.0c08157.

[12] J.B. Goodenough, How we made the Li-ion rechargeable battery, Nat. Electron 1 (2018) 204, https://doi.org/10.1038/s41928-018-0048-6.

[13] D. Deng, Li-ion batteries: basics, progress, and challenges, Energy Sci. Eng. 3 (2015) 385–418, https://doi.org/10.1002/ese3.95.

[14] T.F. Fuller, J.N. Harb, Electrochemical-Engineering, 2018.

[15] J. Christensen, P. Albertus, R.S. Sanchez-Carrera, T. Lohmann, B. Kozinsky, R. Liedtke, J. Ahmed, A. Kojic, A critical review of Li/air batteries, J. Electrochem. Soc. 159 (2011) R1, https://doi.org/10.1149/2.086202jes.

[16] Y. Berro, M. Balat-Pichelin, Metal fuels production for future long-distance transportation through the carbothermal reduction of MgO and Al_2O_3: a review of the solar processes, Energy Convers. Manag. 251 (2022) 114951, https://doi.org/10.1016/j.enconman.2021.114951.

[17] R.B. Subramanyam, Some recent innovations in the Kroll process of titanium sponge production, Bull. Mater. Sci. 16 (1993) 433–451, https://doi.org/10.1007/BF02757646.

[18] K. Araci, D. Mangabhai, K. Akhtar, 9 - production of titanium by the Armstrong Process®, in: M. Qian, F.H. (Sam, Froes (Eds.), Titanium Powder Metallurgy, Butterworth-Heinemann, Boston, 2015, pp. 149–162, https://doi.org/10.1016/B978-0-12-800054-0.00009-5.

[19] W.E. Haupin, Electrochemistry of the Hall-Heroult process for aluminum smelting, J. Chem. Educ. 60 (1983) 279, https://doi.org/10.1021/ed060p279.

[20] G.Z. Chen, D.J. Fray, T.W. Farthing, Direct electrochemical reduction of titanium dioxide to titanium in molten calcium chloride, Nature 407 (2000) 361–364, https://doi.org/10.1038/35030069.

[21] M. Ma, D. Wang, W. Wang, X. Hu, X. Jin, G.Z. Chen, Extraction of titanium from different titania precursors by the FFC Cambridge process, J. Alloys Compd. 420 (2006) 37–45, https://doi.org/10.1016/j.jallcom.2005.10.048.

[22] A.P. Ratvik, R. Mollaabbasi, H. Alamdari, Aluminium production process: from Hall–Héroult to modern smelters, ChemTexts 8 (2022) 10, https://doi.org/10.1007/s40828-022-00162-5.

[23] A.E. Saba, A.E. Elsherief, Continuous electrowinning of zinc, Hydrometallurgy 54 (2000) 91–106, https://doi.org/10.1016/S0304-386X(99)00061-4.

[24] J.M. Bergthorson, Y. Yavor, J. Palecka, W. Georges, M. Soo, J. Vickery, S. Goroshin, D.L. Frost, A.J. Higgins, Metal-water combustion for clean propulsion and power generation, Appl. Energy 186 (2017) 13–27, https://doi.org/10.1016/j.apenergy.2016.10.033.

[25] K. Wefers, C. Misra, P.R. Bridenbaugh, Oxides and Hydroxides of Aluminum, 2003. https://api.semanticscholar.org/CorpusID:106399749.

CHAPTER TWO

Activation of metal fuels

2.1 Overview

Chapter 2 covers the activation of metal fuels, building on the previous chapter's discussion on how metal fuels work. Metal fuels in powder form must be activated to react/burn with oxidizers like H_2O and O_2 quickly. Generally, the rate at which a fuel reacts/burns is determined by its physical state, whether it is a gas, liquid, or solid fuel. Since gases always tend to mix, gas fuels such as natural gas, propane, and hydrogen easily mix with oxidizers in gas forms like O_2 gas and steam to create a homogenous gas–gas mixture. The high surface area of gas molecules in this mixture allows gas fuels to react quickly and efficiently, making gas fuels burn faster than liquid and solid fuels [1,2]. On the other hand, liquid fuels such as gasoline, oil, and liquefied H_2 burn at a moderate rate depending on their volatility. Volatile liquid fuels like gasoline quickly vaporize and mix with O_2 to enable faster burning.

In contrast, heavier liquid fuels like oil must first break into fine droplets to increase surface area and facilitate faster burning. Due to these intermediate steps (i.e., evaporation and breaking into finer particles to form a combustible mixture with O_2), the combustion of liquid fuels is slower than that of gas fuels [1]. Finally, solid fuels such as coal, wood, and metals tend to burn at a slower rate compared to gas and liquid fuels. This is because the combustion of solids often involves multiple stages, including heat absorption by the solid to reach its ignition temperature and the decomposition of this solid into volatile gases that can mix with O_2 and burn. Thus, the combustion rate of solid fuels is limited by the rate at which heat is transferred into the solid interior and volatile gases are released from the solid surface [3]. Due to these multiple stages, the combustion rate of solid fuels is slower than that of gas and liquid fuels.

Considering the abovementioned factors, it becomes clear that despite their high energy density and potential to generate energy without CO_2 emissions, metal fuels in the bulk form fundamentally suffer from a poor oxidation/burning rate, as illustrated in Fig. 2.1 for the $Zn-H_2O$ reaction (i.e., wet oxidation of Zn fuel) introduced in the previous chapter: the

Metals as Clean Fuels
ISBN: 978-0-443-13537-8
https://doi.org/10.1016/B978-0-443-13537-8.00010-5

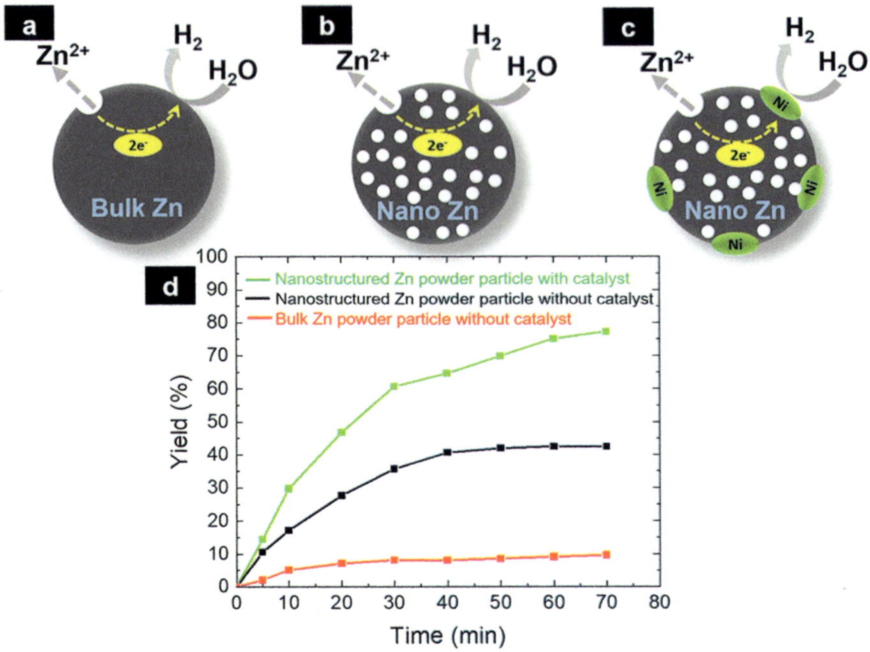

Figure 2.1 The Zn—H_2O reaction is used as a model reaction to illustrate the need to activate metal fuels. (a,d) A dense Zn powder particle gives a lower yield upon reaction with H_2O to produce heat and H_2. (b,d) A nanoporous Zn powder particle gives a higher yield upon reaction with H_2O to produce heat and H_2. (c,d) Decorating the surface of nanoporous Zn with Ni catalyst further increases the reaction yield significantly.

wet oxidation of a dense Zn particle (Fig. 2.1a) results in a lower reaction yield (red curve Fig. 2.1d), while the wet oxidation of a nanoporous Zn particle (Fig. 2.1b) gives a much higher reaction yield (black curve Fig. 2.1d). Therefore, to enhance the reaction rate of metal fuels, it is necessary to use metal fuel particles with *nanoscale features* to increase the contact area between the metal and the oxidizer. Note, however, that using metal particles with nanoscale features alone does not significantly increase the reaction rate. This is because even in the nanostructured format, bulk metal fuels are still intrinsically denser than other forms of solid fuels, such as coal and wood, which makes it challenging to ignite and burn metal fuels.

To significantly boost the reaction rate of metal fuels, it is essential to combine nanostructuring with chemical reaction promoters like catalysts, as illustrated in Fig. 2.1c and d, and physical reaction promoters like high temperature and pressure. This justifies the importance of activating metal fuels. There are various approaches for activating metal fuels, which can be grouped into three main methods discussed further in this chapter: chemical activation using catalysts and reactor promoters, activation through nanostructuring, which is the focus of this monograph, and finally, thermal activation.

2.2 Chemical activation of metal fuels using catalysts

Metal fuels can be chemically activated using a solid or liquid chemical agent, depending on the nature of the oxidizer. These chemical agents can be classified into two groups. The first group corresponds to catalysts, which improve the reaction rate by lowering the activation energy barrier of the reaction. The second group corresponds to reaction promoters. This section focuses on catalysts; reaction promoters will be discussed in the next section.

To illustrate how a catalyst works during metal fuel oxidation, we will examine the reaction between Zn and H_2O as an example of wet metal fuel oxidation. Let's look closely at the Zn—H_2O reaction discussed in Chapter 1, which is summarized by Eq. (1.12). This reaction is an oxidation-reduction (redox) reaction since one of the reactants, Zn, gets oxidized (i.e., the oxidation number of Zn changes from 0 in metallic Zn before the reaction to +2 in ZnO after the reaction), while the other reactant, H_2O, gets reduced (i.e., the oxidation number of the hydrogen atom changes from +1 in H_2O before the reaction to 0 in H_2 after the reaction). Hence, we can break down the redox reaction in Eq. (1.12) into an oxidation half-reaction given in Eq. (2.1) and a reduction half-reaction in Eq. (2.2). The overall Zn—H_2O reaction in Eq. (1.12) can only proceed if all the electrons produced in the oxidation half-reaction (Eq. 2.1) are consumed in the reduction half-reaction (Eq. 2.2). In other words, the rate at which electrons are released during Zn oxidation should equal the rate at which electrons are consumed when H_2O is reduced.

It is important to note that for H_2O to receive electrons from the oxidizing metal, the H_2O molecule should be in physical contact with the metal surface. Additionally, some metals catalyze the reduction of H_2O to H_2 and hydroxide, while others do not. Therefore, the rate at which electrons are consumed in the reduction half-reaction in Eq. (2.2), and ultimately the overall reaction rate of the metal-H_2O reaction in Eq. (1.12), depends on the ability of the metal surface to catalyze the reduction of H_2O to H_2 and hydroxide. Because metallic Zn is intrinsically a very poor H_2 evolution catalyst, in this chapter, we purposely selected the Zn—H_2O reaction in Eq. (1.12) as an example of wet metal fuel oxidation that requires a catalyst to enhance the reaction rate. Thus, in the absence of a catalyst, the rate at which H_2O is reduced on the surface of Zn to form H_2 and hydroxide through the half-reaction in Eq. (2.2) is relatively low, even when nanostructured Zn is used, as illustrated in Fig. 2.1b and d. On the other hand, since Ni is an effective H_2 evolution catalyst in neutral and alkaline solutions, by decorating the surface of nanostructured Zn with a small amount of Ni atom clusters used as an H_2 evolution catalyst, the H_2O reduction half-reaction in Eq. (2.2) will occur preferentially on the Ni atom clusters and at a higher rate, as illustrated in Fig. 2.1c and d. This will boost the overall rate of the Zn—H_2O reaction in Eq. (1.12).

$$\text{Oxidation half} - \text{reaction: Zn (s)} + 2\text{OH}^-(\text{aq}) \rightarrow \text{ZnO(s)} + \text{H}_2\text{O(l)} + 2\text{e}^- \tag{2.1}$$

$$\text{Reduction half} - \text{reaction: } 2\text{H}_2\text{O(l)} + 2\text{e}^- \rightarrow \text{H}_2(\text{g}) + 2\text{OH}^-(\text{aq}) \tag{2.2}$$

$$\text{Total reaction: } \overline{\text{Zn (s)} + \text{H}_2\text{O(l)} \rightarrow \text{ZnO(s)} + \text{H}_2(\text{g})} \tag{1.12}$$

Drawbacks of catalysts: While catalysts can effectively enhance the reaction rate during metal fuel oxidation, for scalable applications, Earth-abundant nonprecious metal catalysts are preferred to costly noble metal catalysts. However, it should be emphasized that under extreme reaction environments, nonprecious metal catalysts can oxidize or corrode.

2.3 Chemical activation of metal fuels using reaction promoters

Besides catalysts, reaction promoters represent another group of chemical agents that can be used to activate metal fuels. Reaction promoters prevent the oxide or hydroxide solid by-products formed during wet metal fuel oxidation from becoming self-limiting. Specifically, metal-H_2O reactions, including the Al—H_2O, Fe—H_2O, and Zn—H_2O reactions discussed in Chapter 1, are self-limiting reactions. This means that once the reaction starts, the oxide or hydroxide layer, such as $Al(OH)_3$, AlOOH, Al_2O_3, ZnO, Fe_3O_4, and Fe_2O_3 coproduced with H_2 and heat, often forms a shell around the corresponding metal core. This shell prevents H_2O from coming into direct contact with the metal underneath to react further. As a result, the reaction stops. Such a core-shell structure is illustrated in Fig. 2.2a for the Al—

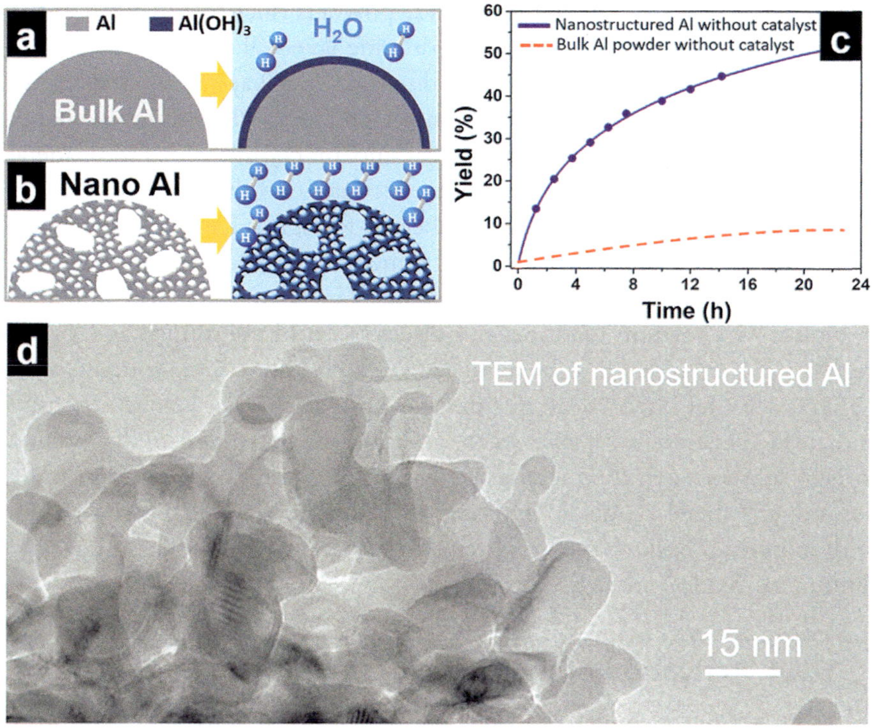

Figure 2.2 (a,c) The reaction of dense Al (gray) with water to produce H_2 is impeded by the self-limiting $Al(OH)_3$ shell (dark blue) solid by-product. (b,c) The reaction of ultrafine nanoporous Al with water is not impeded because, by the time a self-limiting $Al(OH)_3$ shell forms, all of the nanostructured Al has reacted. (d) Typical TEM image of ultrafine nanoporous Al with very small feature size in the range of 10–15 nm.

H_2O reaction. The most popular solution to overcome this self-limiting oxide or hydroxide layer is to modify the chemical composition of the reactive metal or the chemical composition of H_2O, by adding reaction promoters in the metal and/or in H_2O [4–12]. One approach to modifying the composition of the reactive metal is to make metal-salt composites that are more reactive with water than pure metals [8]. Another approach is to make alloys by adding a small amount of metallic elements, such as Ga, Li, Sn, In, and Co, to the base metal. The corresponding alloys are then more water-reactive than pure metals. For example, in the case of the Al–H_2O reaction, alloys of Al with liquid Ga and Al with eutectic liquid Ga-In-Sn are very effective in suppressing the Al oxide/hydroxide layer by continuously enabling a fresh Al surface to react with H_2O [13]. The modification of the composition of H_2O is typically done by adding specific salts (such as

NaCl, NaOH) or metal oxide/hydroxide powders (such as $Al(OH)_3$) in the H_2O solution to disrupt the oxide/hydroxide layer [9].

Drawbacks of reaction promoters: As discussed in Chapter 1, what makes metal fuels attractive is their very high gravimetric and volumetric energy densities, defined as the amount of heat and/or H_2 released during metal fuel oxidation divided by the stoichiometric mass or volume of all the reactants. Therefore, adding a large number of foreign substances as reaction promoters to the metal-H_2O system to remove the oxide or hydroxide layer is not an ideal solution for two main reasons. Firstly, any substance added to the metal-H_2O system that does not directly contribute to heat and/or H_2 generation is considered to be a "dead" mass (volume). Consequently, this will reduce the theoretical gravimetric and volumetric energy density and/or H_2 generation capacity of the metal-H_2O system. Secondly, adding foreign substances to the metal-H_2O system will result in impurities in the coproduced metal oxide or hydroxide. Reducing/recycling metal oxides with impurities back to the underlying metals is more costly than recycling pure metal oxides.

2.4 Activation of metal fuels by nanostructuring

The rate of metal-O_2 and metal-H_2O reactions is significantly impacted by the size of the metal. For example, as discussed earlier for metal-H_2O reactions, the reaction between metals and H_2O does not continuously proceed because the oxide or hydroxide layer formed around the metal prevents H_2O from coming into direct contact with the metal underneath. This phenomenon is illustrated in Fig. 2.2a, where the dark blue shell representing $Al(OH)_3$ covers the bulk Al metal in gray [14]. Instead of suppressing this oxide/hydroxide layer by chemically modifying the composition of the metal-H_2O system using reaction promoters, as discussed in Section 2.3, another way to overcome this issue is to modify the microstructure of the metal through nanostructuring. Let's look closely at the Al$-$$H_2O$ reaction (Fig. 2.2a and b).

It is important to note that this reaction can only proceed fully without reaction promoters if the size of the Al metal is around 7 nm [14]. This is because 7 nm is approximately twice the thickness of the blocking Al oxide/hydroxide layer (3$-$4 nm). Under these conditions, during the reaction of Al with H_2O to release H_2 and coproduce $Al(OH)_3$, by the time a self-limiting oxide/hydroxide layer fully forms around the Al nanoparticle, all of the Al has reacted with H_2O, as illustrated in Fig. 2.2b. Hence, the activation of metal fuels can be achieved by reducing the size of the metal from the macroscale range down to a certain critical size in the nanoscale range (this critical size is approximately 7 nm for Al). Doing so will drastically enhance the rate of metal-H_2O and metal-O_2 reactions.

Metal fuel activation through nanostructuring can be achieved by various methods. One such method is mechanical, which involves using ball-milling techniques to break down bulk metals into smaller pieces until they reach the nanoscale. Ball milling can be done at near room temperatures or under cryogenic conditions [15]. Another method is chemical or electrolytic, which involves using selective etching or dealloying to chemically or electrolytically remove the most reactive component of an alloy and create nanoporous metals [16]. In this monograph, we will primarily focus on the latter method, i.e., creating nanoporous metal fuels through selective chemical or electrolytic dealloying. In Chapter 3, we will introduce and discuss the dealloying process in detail. In Chapters 4 and 5, we will discuss the fabrication of nanoporous Zn fuel by chemical etching. As a glimpse, the nanoporous Zn powder shown in Fig. 2.3 was made by chemical etching of Al from Zn—Al parent alloys.

Chapters 6 and 7 will cover the fabrication of nanoporous Al using electrolytic dealloying. As a glimpse, the nanoporous Al shown in Fig. 2.2d was created through selective electrolytic etching of Mg from Al—Mg parent alloys. Until now, the major limitation of the selective etching process is that the sacrificial metals etched from the parent alloy, which often account for over 50% of the parent material, usually end up as waste products. Etching without sacrificial material loss is a crucial step toward a sustainable, circular economy. Therefore, in Chapters 8 and 9, we will demonstrate a novel, versatile electrolytic method for making nanoporous metals that nearly eliminates the loss of sacrificial materials. This method, which we call chlorine-neutral etching, uses Cl to introduce nanoporosity in metals through the Kirkendall effect, with nearly complete recovery of Cl. We will showcase this method by making noble nanoporous metal/metal composites, including nanoporous Ag/Ag composites (Chapter 8) and nanoporous Ag/Au composites (Chapter 9), with the hope that this method could be extended in the future to make nonprecious nanoporous metal/metal composites.

Drawbacks of nanostructuring: In general, processes to make nanostructured materials are not always scalable. For example, the ball-milling process to make nanostructured metals is highly energy-intensive, particularly cryogenic milling. Such methods can only be used to create a relatively small

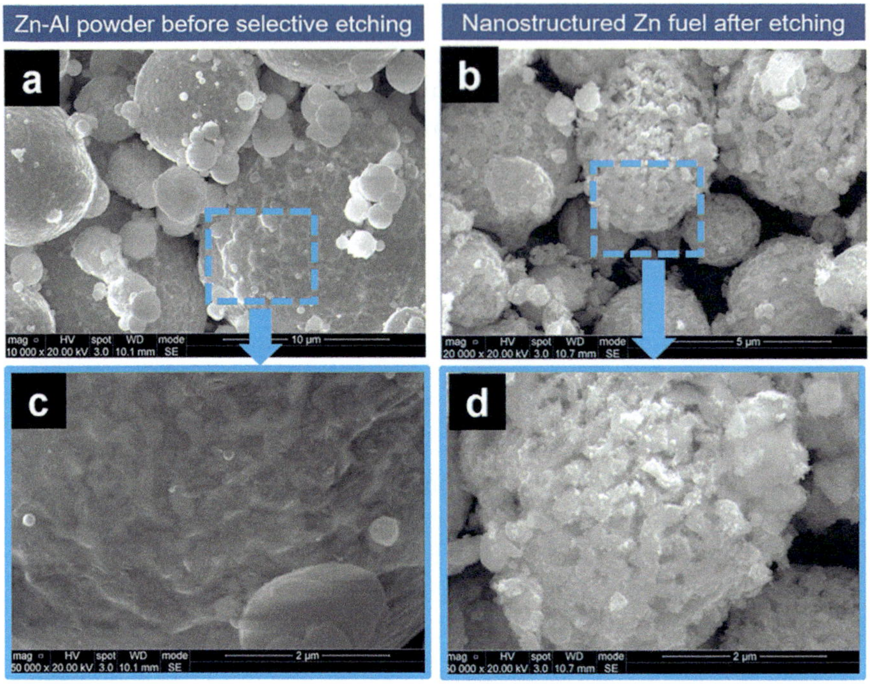

Figure 2.3 (a,c) Micro-sized gas-atomized Zn—Al parent alloy powder particles at low (a) and (c) magnification before chemical dealloying. (b,d) Corresponding nanoporous Zn powder particles at low (b) and (d) magnifications after chemical dealloying.

amount of nanostructured metals at the laboratory scale or for specific applications. They are not suitable for large-scale synthesis. Selective etching, on the other hand, can be used to make nanostructured metals for large-scale applications. For example, nanoporous nickel, commonly known as *Raney nickel* and made by selectively etching Al from Ni—Al alloys, is an essential catalyst in the industry for hydrogenating benzene to cyclohexane [17,18]. The main drawbacks of the selective etching process are (i) the slow etching rate, which can be addressed using microsized parent alloys, as will be discussed in this monograph, and (ii) the loss of sacrificial materials during dealloying, which in the case of electrolytic dealloying can be addressed by recovering the sacrificial component during dealloying, as will be discussed in this monograph. The loss of sacrificial elements can also be addressed using the novel chlorine-neutral etching methods that we will introduce in Chapters 8 and 9. Another drawback of nanostructured metals is their high reactivity and pyrophoricity, which may pose safety concerns.

The monograph concludes with the characterization of the fascinating intertwining of two different items in porous structures, i.e., geometry and topology. In Chapter 10 we discuss a theoretical approach derived from integral-geometry-based morphological image analysis of highly porous structures where effects of geometry and topology are tangling and intertwisting. It is essential when characterizing the structure—property relationships of porous materials to have quantitative numbers about the connectivity in terms of geometry and topology. As emphasized in Chapter 10, the essential and appealing feature of the integral-geometry-based approach is that the complexity of the mathematical framework is in no way comparable to the simplicity of its practical application. These theoretical concepts are confronted with various porous structures, and in particular, it will be shown that the methodology yields a quantitative characterization of two- and three-dimensional patterns in terms of geometrical and topological descriptors called Minkowski functionals.

2.5 Thermal activation of metal fuels and their drawbacks

The rate of chemical reactions exponentially increases with increasing temperature in accordance with the Arrhenius relation in Eq. (2.3):

$$k \propto \exp\left(-\frac{E}{k_B T}\right) \tag{2.3}$$

where k, E, T and k_B represent the rate constant, activation energy, temperature, and Boltzmann constant, respectively. As such, an effective way to boost the rate of metal-H_2O reactions is by increasing the reaction temperature. However, since metal-H_2O reactions are exothermic, moderate metal-H_2O reaction temperatures can be achieved without the need to supply external heat, simply by taking advantage of the heat released from these exothermic reactions. Thus, performing metal-H_2O reactions at high temperatures significantly enhances the efficiency of the reaction. Another advantage of these high-temperature reactions is that the Gibbs free energy decreases as the temperature increases. This means that the input energy required to drive a metal-H_2O reaction will decrease with higher temperatures. In fact, this principle justifies why high-temperature electrolysis of H_2O vapor is less energy-intensive compared to room-temperature electrolysis of H_2O liquid. On the other hand, processes taking place at high temperatures and pressures are not always practical, and represents the main drawback of this activation approach.

References

[1] Y.-H. Kiang, Chapter 9 - Combustion fundamentals and energy systems, in: Y.-H. Kiang (Ed.), Fuel Property Estimation and Combustion Process Characterization, Academic Press, 2018, pp. 313–362, https://doi.org/10.1016/B978-0-12-813473-3.00009-X.
[2] L. Zhou, Theory and Modeling of Dispersed Multiphase Turbulent Reacting Flows, Butterworth-Heinemann, Oxford, England, 2018.
[3] M. Kumar, S. Kumar, S.K. Tyagi, Design, development and technological advancement in the biomass cookstoves: a review, Renew. Sustain. Energy Rev. 26 (2013) 265–285, https://doi.org/10.1016/j.rser.2013.05.010.
[4] H.Z. Wang, D.Y.C. Leung, M.K.H. Leung, M. Ni, A review on hydrogen production using aluminum and aluminum alloys, Renew. Sustain. Energy Rev. 13 (2009) 845–853, https://doi.org/10.1016/j.rser.2008.02.009.
[5] A. V Ilyukhina, A.S. Ilyukhin, E.I. Shkolnikov, Hydrogen generation from water by means of activated aluminum, Int. J. Hydrogen Energy 37 (2012) 16382–16387, https://doi.org/10.1016/j.ijhydene.2012.02.175.
[6] L. Soler, A.M. Candela, J. Macanás, M. Muñoz, J. Casado, In situ generation of hydrogen from water by aluminum corrosion in solutions of sodium aluminate, J. Power Sources 192 (2009) 21–26, https://doi.org/10.1016/j.jpowsour.2008.11.009.
[7] L. Soler, A.M. Candela, J. Macanás, M. Muñoz, J. Casado, Hydrogen generation from water and aluminum promoted by sodium stannate, Int. J. Hydrogen Energy 35 (2010) 1038–1048, https://doi.org/10.1016/j.ijhydene.2009.11.065.
[8] G.L. Ma, H. Bin Dai, D.W. Zhuang, H.J. Xia, P. Wang, Controlled hydrogen generation by reaction of aluminum/sodium hydroxide/sodium stannate solid mixture with water, Int. J. Hydrogen Energy 37 (2012) 5811–5816, https://doi.org/10.1016/j.ijhydene.2011.12.157.
[9] Y.-K. Chen, H.-T. Teng, T.-Y. Lee, H.-W. Wang, Rapid hydrogen generation from aluminum–water system by adjusting water ratio to various aluminum/aluminum hydroxide, Int. J. Energy Environ. Eng. 5 (2014) 87, https://doi.org/10.1007/s40095-014-0087-3.
[10] V. Rosenband, A. Gany, Application of activated aluminum powder for generation of hydrogen from water, Int. J. Hydrogen Energy 35 (2010) 10898–10904, https://doi.org/10.1016/j.ijhydene.2010.07.019.
[11] X. Huang, C. Lv, Y. Huang, S. Liu, C. Wang, D. Chen, Effects of amalgam on hydrogen generation by hydrolysis of aluminum with water, Int. J. Hydrogen Energy 36 (2011) 15119–15124, https://doi.org/10.1016/j.ijhydene.2011.08.073.
[12] Z. Zhao, X. Chen, M. Hao, Hydrogen generation by splitting water with Al-Ca alloy, Energy 36 (2011) 2782–2787, https://doi.org/10.1016/j.energy.2011.02.018.
[13] J.T. Ziebarth, J.M. Woodall, R.A. Kramer, G. Choi, Liquid phase-enabled reaction of Al–Ga and Al–Ga–In–Sn alloys with water, Int. J. Hydrogen Energy 36 (2011) 5271–5279, https://doi.org/10.1016/j.ijhydene.2011.01.127.
[14] J.S. Corsi, J. Fu, Z. Wang, T. Lee, A.K. Ng, E. Detsi, Hierarchical bulk nanoporous aluminum for on-site generation of hydrogen by hydrolysis in pure water and combustion of solid fuels, ACS Sustain. Chem. Eng. 7 (2019) 11194–11204, https://doi.org/10.1021/acssuschemeng.9b00481.
[15] S.S. Razavi-Tousi, J.A. Szpunar, Effect of structural evolution of aluminum powder during ball milling on hydrogen generation in aluminum–water reaction, Int. J.

Hydrogen Energy 38 (2013) 795–806, https://doi.org/10.1016/j.ijhydene.2012.10.106.

[16] J. Fu, S.S. Welborn, E. Detsi, Dealloyed air- and water-sensitive nanoporous metals and metalloids for emerging energy applications, ACS Appl. Energy Mater. 5 (2022) 6516–6544, https://doi.org/10.1021/acsaem.2c00405.

[17] B. Zeifert, J.S. Blásquez, J.G.C. Moreno, H.A. Calderón, Raney-nickel catalysts produced by mechanical alloying, Rev. Adv. Mater. Sci. 18 (2008) 632–638.

[18] T.W.G. Solomons, C.B. Fryhle, Organic Chemistry, 2012, https://doi.org/10.1007/s13398-014-0173-7.2.

CHAPTER THREE

Fundamentals of dealloying

3.1 Overview

Dealloying is broadly defined as the selective removal of the most chemically or thermally active element or phase from a dense multicomponent parent alloy. Dealloying can be done by chemical etching or thermal evaporation, resulting in the formation of a bicontinuous nanoporous architecture, as shown in Fig. 3.1a and b in the case of nanoporous gold and Fig. 3.1c in the case of nanoporous aluminum. Such a *bicontinuous* nanoporous architecture is characterized by the coexistence of a continuous network of nanoscale solid structures, called ligaments, and a continuous network of open nanochannels, referred to as pores [1]. Dealloying, or selective alloy etching, has been used since ancient civilizations. For example, depletion gilding, also known as depletion plating, is a variation of selective alloy etching practiced since the Bronze Age to alter the appearance of metallic objects by depleting Cu near the surface of utensils and jewelry made of Au-Cu and Ag-Cu alloys, making them look like pure Au and pure Ag (also known as Sterling Ag), respectively [2].

In modern days, selective alloy etching has been widely used to create bicontinuous nanoporous metals for various applications, including (electro)catalysis and energy conversion and storage [1,3,4]. For example, nanoporous nickel, commonly known as *Raney nickel* and made by selectively etching Al from Ni-Al alloys, is an essential catalyst in the industry for hydrogenating benzene to cyclohexane [5,6]. More generally, nanoporous metals and metalloids are very attractive for applications such as heterogeneous catalysis [7—9], electrocatalysis [10—13], electrochemical energy

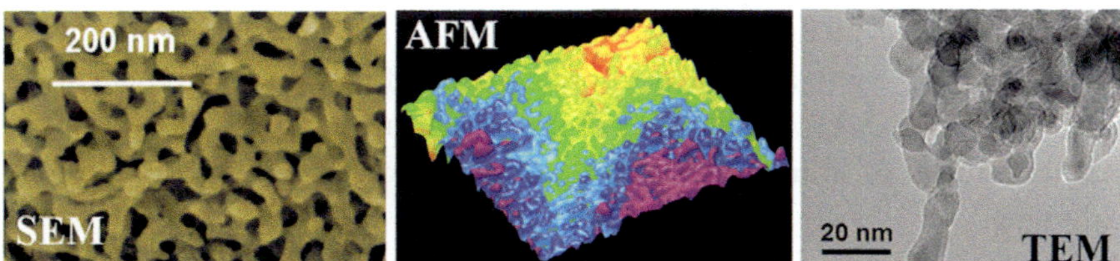

Figure 3.1 Left: SEM of NP-Au; middle: AFM of NP-Au; right: TEM of NP-Al.

Metals as Clean Fuels
ISBN: 978-0-443-13537-8
https://doi.org/10.1016/B978-0-443-13537-8.00007-5

conversion and storage [3,14–22], electrochemical actuators [23–26], sensors [27–29], surface-enhanced Raman scattering (SERS) [30–32], and plasmonics [33–35]. For this reason, to date, a broad range of nanoporous metals and metalloids, including nanoporous Au (Fig. 3.1a and b), Pt, Pd, Ag, Cu, Sn, Ni, Co, Si, Ge, B, Sb, Pb, Ti, Mo, Nb, Al (Fig. 3.1c), Mg, and Zn, have been fabricated using various dealloying strategies depending on the chemical and thermal reactivity of the elements involved.

As mentioned in the previous chapter, metal fuels in bulk form exhibit a poor reaction rate during dry and wet oxidation. As such, metal fuels need to be activated to enhance their reactivity. The main focus of this monograph is to demonstrate a scalable dealloying approach to activating metal fuels by introducing nanoporosity within these metals. The use of some dealloyed nanoporous metals in industry (i.e., Raney nickel) clearly suggests that it should be possible to develop a scalable dealloying process to activate metal fuels for large-scale applications. To bring the reader up to speed with the subject, this chapter is aimed at elucidating the fundamental mechanism that governs the formation of nanoporosity via a spinodal decomposition pathway during dealloying and elucidating the chemical and electrochemical reaction mechanisms associated with state-of-the-art dealloying methods developed by the community.

3.2 Nanoporosity formation via a spinodal decomposition pathway

Spinodal decomposition is the spontaneous separation of a single phase (purple color, Fig. 3.2) into two new phases (blue and red colors, Fig. 3.2) without nucleation. In the case of solid-to-solid decomposition, during which a single solid phase separates into two new solid phases, all those phases have the same crystal structure but different chemical compositions. A key characteristic of a spinodal-decomposed structure is the *bicontinuous* nature of the two new phases. Interestingly, the SEM image of nanoporous gold in Fig. 3.1a is very similar to that of a spinodal-decomposed structure consisting of two phases, a solid phase made of interconnected ligaments (gold color, Fig. 3.1a), and an open phase made of interconnected channels (black color, Fig. 3.1a). Therefore, an analogy can be investigated between the mechanisms of spinodal decomposition and nanoporosity formation by dealloying.

In the discussion of spinodal decomposition, we consider the hypothetical binary alloy phase diagram in Fig. 3.3a, which consists of two elements,

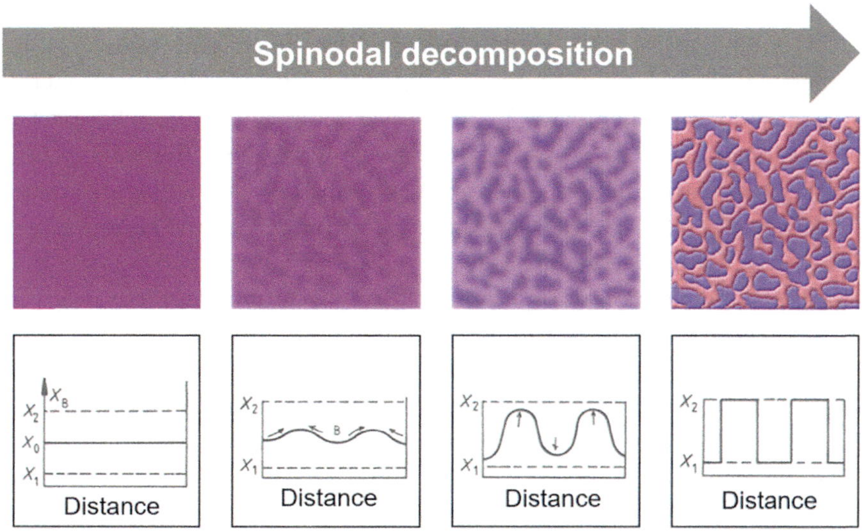

Figure 3.2 Upper row: Morphology evolution during spinodal decomposition in an alloy quenched into the spinodal region. Lower row: Corresponding evolution of the composition profile with increasing time. *The lower row illustrations are adapted from D.A. Porter, K. E. Easterling, M.Y. Sherif, Phase Transformations in Metals and Alloys, third ed., (Chapter 5, Fig. 5.39).*

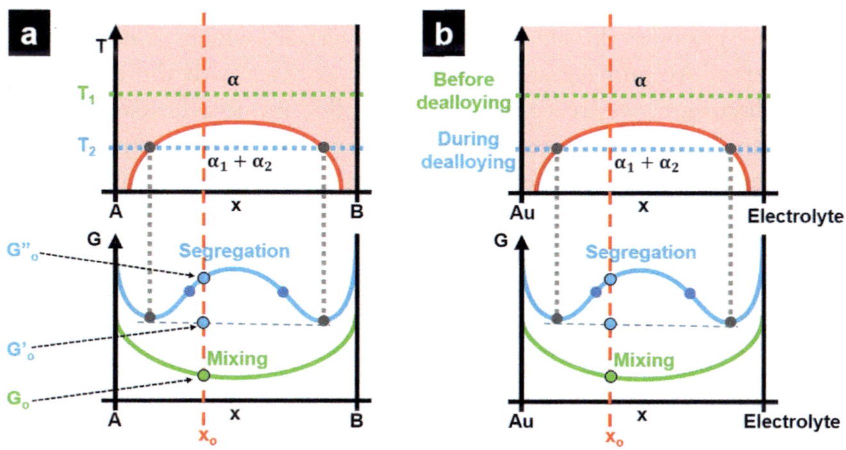

Figure 3.3 Schematic representation of the phase diagram (upper schema) and corresponding Gibbs free energy versus composition curve (lower schema) for (a) an Au-Ag and (b) Au-electrolyte regular solution mixing (green lines) and segregation (blue lines).

A with phase α_1 and B with phase α_2. The two elements, A and B, do not mix well at low temperatures (mixability gap), but they mix very well at high temperatures to form a single phase, α. If an alloy with composition X_o (red vertical line Fig. 3.3) within the spinodal points is aged at a high-temperature T_1, where it exists as a single phase α with Gibbs free energy G_o, and suddenly taken from T_1 to a lower temperature T_2, where it has the choice to exist either as a single phase α with Gibbs free energy G''_o or to separate into a mixture of two phases α_1 and α_2 with Gibbs free energy G'_o (where $G'_o < G''_o$), then this alloy will naturally take the configuration with lower Gibbs free energy G'_o by separating into α_1 and α_2. In other words, phase separation during spinodal decomposition (Fig. 3.2) is driven by free energy minimization.

To help understand why the morphology of dealloyed nanoporous metals (Fig. 3.1) is similar to that of spinodal-decomposed structures (Fig. 3.2), an analogy can be drawn between the mechanisms of dealloying and spinodal decomposition. We will use the formation of nanoporous gold by selective removal of silver from a gold-silver (Au-Ag) alloy as a model system for the purpose of this discussion. Nanoporous gold (Fig. 3.1a and b) is commonly fabricated by selective removal of Ag from Au-Ag alloys with a typical composition around $Au_{35}Ag_{65}$ at.%. Fig. 3.4 (top) represents an $Au_{35}Ag_{65}$ parent alloy submerged in a corroding solution or electrolyte, with yellow and grey marks representing gold and silver atoms, respectively. The light blue region represents the bulk of the corroding electrolyte. The darker blue region is the electrolyte/alloy interface, which contains only a monolayer of Au and Ag atoms that are in contact with the electrolyte. For this discussion, we treat this electrolyte/alloy monolayer (dark blue region) as an "alloy" composed of the two components: (i) remaining Au atoms after Ag atoms have been etched away from this Au-Ag monolayer; and (ii) the corroding electrolyte. This framework will be used to explain the spinodal decomposition pathway in dealloying.

(i) *Single-phase system (mixing):* Upon submersion of the Au-Ag alloy in the corroding electrolyte, Ag atoms initially dissolve from the Au-Ag monolayer at the electrolyte/alloy interface (dark blue region Fig. 3.4—top), leaving behind Au adatoms. In this electrolyte/alloy monolayer system, there exists a solubility limit X_L of Au atoms in the electrolyte [36]. It is relevant to realize that the physical picture presented is fully based on "enthalpy" and the initial atomic flatness with only one single Ag atom missing, sounding rather super ideal, whilst "entropy" effects are fully ignored. If X_o represents the concentration of Au atoms in that monolayer system, as long as $X_o < X_L$ (i.e., the

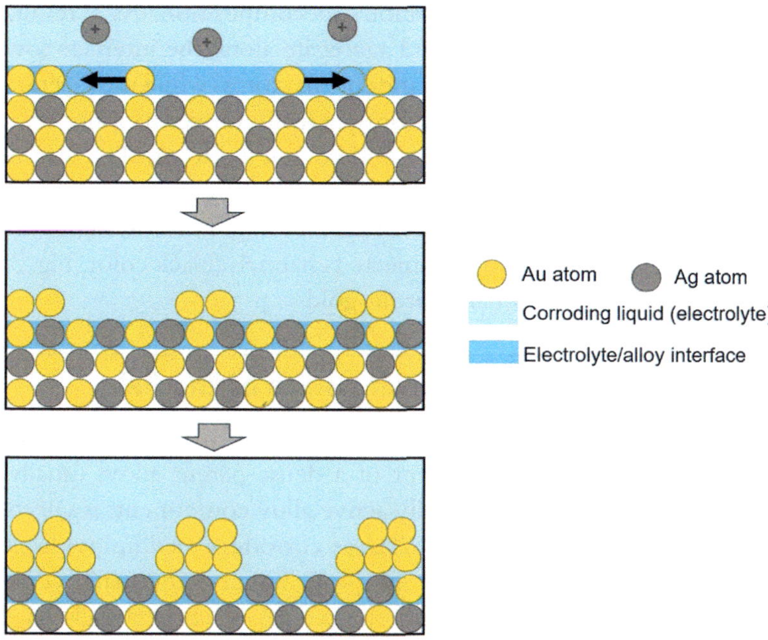

Figure 3.4 Schematic illustration of dealloying and nanoporosity formation through the spinodal decomposition pathway for an Au-Ag parent alloy (Au atoms are in yellow, and Ag atoms are in gray). The Au-Ag alloy is in contact with corroding liquid (light blue). The dark blue monolayer represents the Au-electrolyte regular solution. The dealloying process involves the selective removal of Ag metal and surface diffusion of Au adatoms to form Au clusters, which later evolve to form Au ligaments.

concentration of Au atoms is lower than the solubility limit), the electrolyte layer and Au atoms in it will mix very well, analogous to the single α phase from Fig. 3.3b (green dotted line), with the corresponding free energy versus composition curve illustrated by the solid green line in Fig. 3.3b.

(ii) *Two-phase system (spinodal decomposition):* As more Ag atoms are etched from the electrolyte/alloy monolayer system, the concentration of Au atoms increases until $X_o > X_L$ (i.e., the concentration of Au atoms becomes higher than the solubility limit). This situation is comparable to the miscibility gap in the phase diagram in Fig. 3.3b (blue dotted line), with the corresponding free energy versus composition curve illustrated by the solid blue line in Fig. 3.3b. Because of this miscibility gap, the α single-phase region (Au atoms dissolved in the electrolyte layer) segregates into two individual phases (α_1—Au atoms; α_2—electrolyte layer) in a process that looks like spinodal decomposition. As a result, Au adatoms may diffuse quickly and aggregate along the interface towards nearby Au adatoms, forming islands of Au atoms, while the electrolyte/ alloy monolayer (dark blue region) gradually moves to inner layers in the bulk of the alloy (Fig. 3.4 middle and bottom). The Au atoms islands further grow to form a bicontinuous structure consisting of a solid phase made of interconnected ligaments (gold color, Fig. 3.1a) and an open phase made of interconnected channels (black color, Fig. 3.1a), which we recognize as nanoporous gold.

3.3 Background and current state of the field

As discussed above, during the fabrication of nanoporous metals by dealloying, the sacrificial component of a dense parent alloy, usually the most (electro)chemically or thermally active alloy component, is selectively removed using either an acidic and alkaline corroding medium without the need for an external electrical voltage [4,37,38] or a thermally corrosive environment. The reduction-oxidation (redox) reaction associated with this process is a *spontaneous* reaction. When the redox reaction that drives the removal of the sacrificial component is not spontaneous, an external electrical voltage or current can be used to drive the dissolution of the sacrificial component. In the former case (no external electrical voltage or current), the process is referred to as free corrosion dealloying [4,37,38]. In the latter case (with external electrical voltage or current), the process is referred to as electrolytic dealloying [4,37,38]. Free corrosion and electrolytic dealloying in aqueous solutions have been widely used to fabricate precious nanoporous metals such as nanoporous Au [39,40], Ag [10,41], Pt [42,43], Pd [44,45], and Cu [30,46]. Although nanoporous metals made of precious elements are very promising for various (electro)chemical applications [11,26,47], precious metals are scarce and, therefore, not attractive for large-scale applications. Pt, for example, is used in state-of-the-art fuel cells and metal-air batteries as an electrocatalyst for the oxygen reduction reaction (ORR) [48,49], but its high cost has motivated the community to search for Pt-free ORR electrocatalysts [50].

In recent years, Earth-abundant air- and water-sensitive nanoporous metals and metalloids, such as nanoporous Al, Mg, Zn, Fe, Si, Sb, Sn, etc., have received significant interest for their broad range of energy applications including metal fuels, ignition aids for solid fuel combustion [51—53], active ingredients in water-powered spaceship engines such as

ALICE (Aluminum ICE) rocket propellant [54–56], heterogeneous catalysts, and electrocatalysts for fuel production (e.g., carbon dioxide reduction and water electrolysis) and fuel consumption (e.g., oxygen reduction, hydrogen and alcohol oxidation) [57–59], materials for onboard and onsite hydrogen production to feed hydrogen fuel cells [60–62], and lithium-ion, sodium-ion, and magnesium-ion battery electrodes [63–66]. The conventional dealloying method that uses aqueous media proves difficult to make nanoporous metals from non-precious elements. One reason for this is the relatively high chemical reactivity of non-precious metals with aqueous media, which often results either in the formation of a passive oxide film during dealloying in aqueous alkaline solutions or in the full dissolution of the starting parent alloy during dealloying in acidic media. To mitigate these issues, the community has developed various dealloying strategies, which can be classified into four main groups:

(i) Nanoporous metals with a relatively low air sensitivity and low chemical reactivity, such as nanoporous Au, Pt, Ag, Cu, Sn, and Ni, etc., are commonly fabricated by free corrosion dealloying or electrolytic dealloying in an aqueous media, during which the most chemically reactive component of a parent alloy is selectively etched away using an appropriate aqueous corroding medium in the case of free corrosion dealloying [1,3], or using an external electrical voltage (or current) in the case of aqueous electrolytic dealloying. Details on the underlying chemical and electrochemical reaction mechanisms will be discussed in Sections 3.4.1–3.4.4.

(ii) Nonprecious nanoporous metals made of refractory elements with relatively high chemical reactivity, such as nanoporous Ti, Mo, and Nb, etc., are often fabricated through a *non-aqueous* free corrosion dealloying process in which a metallic melt is used as the corroding medium, instead of an aqueous corroding medium. The use of molten metal as the corroding medium is based on the difference in enthalpy of mixing between the underlying alloy elements and the molten metal [67]. Further details on this reaction mechanism will be discussed in Section 3.4.9.

(iii) Nanoporous metals and metalloids can be produced through vapor phase dealloying. Having a parent alloy composed of two components, one with a high melting point and low vapor pressure and the other with a relatively low melting point and high vapor pressure, the sacrificial component with a low melting point and high vapor pressure can be selectively removed by thermal evaporation, leaving behind the nanoporous structure [68]. This process will be further discussed in Section 3.4.4.

(iv) The fabrication of nonprecious nanoporous metals without surface oxide coverage, using nonrefractory metals with high chemical reactivity, such as nanoporous Al and nanoporous Mg, has been a significant challenge in the field. One of the reasons for this challenge is the high chemical reactivity of this class of nonprecious metals, which makes it difficult to use aqueous electrolytes as corroding media. The other reason is the relatively low melting point and high vapor pressure (particularly in the case of Mg) of these metals, which also makes it difficult to use thermal evaporation to selectively evaporate the sacrificial component. A promising solution to overcome these issues is to carry out the dealloying process in an inert environment using anhydrous organic electrolytes. Significant efforts have been made in the field on the fabrication of nanoporous metals in nonaqueous organic electrolytes: (1) Sieradzki and Chen demonstrated the fabrication of nanoporous Sn and Pb by electrolytic dealloying in organic electrolytes with Li ion as the working ion [69]. (2) We demonstrated the fabrication of nanostructured Sn by electrolytic dealloying of Mg_2Sn (i.e., $Sn_{33}Mg_{67}$ at.%) parent alloys in anhydrous organic electrolytes with Mg ion as the working ion [63]. (3) We have also demonstrated the fabrication of ultrafine and oxide-free nanoporous Al by electrolytic dealloying of Al-Mg alloys in anhydrous organic electrolytes with Mg ion as the working ion [15], and ultrafine nanoporous Mg by electrolytic dealloying of Mg-Li alloys in anhydrous electrolytes with Li ion as the working ion [70].

Thanks to the various efforts in the field, briefly introduced in (i) to (iv), nowadays, dealloying can be used to make nanoporous materials from nearly any practical element in the periodic table. The main focus of this monograph is on air- and water-sensitive nanoporous metals, which are more suitable for use as fuels than precious metals.

First, we provide three key selection criteria for the air- and water-sensitive elements, highlighted in red and green in the periodic table in Fig. 3.5. The first criterion is the reactivity with water. Air- and water-sensitive materials of interest are elements with standard reduction potentials below 0 V versus the standard hydrogen electrode (SHE) [71]. One exception is Sb, which has a slightly positive standard reduction potential ($E_O = +0.214$ V vs. SHE for the Sb^{3+}/Sb reaction) [72]. Elements marked in green in Fig. 3.5 are water-reactive but can, in principle, be fabricated in

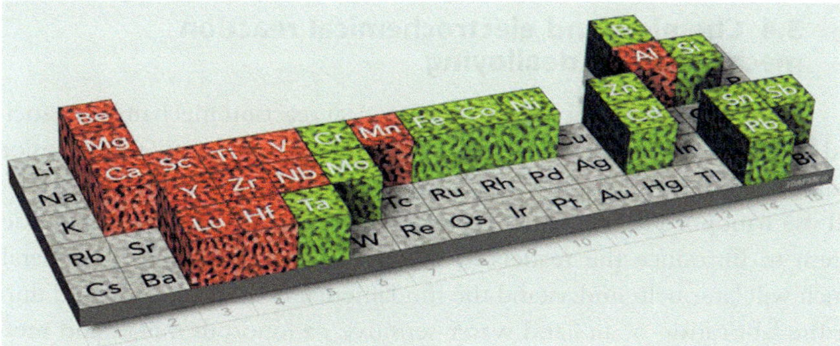

Figure 3.5 Air- and water-sensitive elements of interest. Elements that can, in principle, be fabricated in aqueous solutions are highlighted in green, and elements that cannot be fabricated in aqueous solutions due to their very high reactivity are highlighted in red.

aqueous solutions with careful control of the pH of the corroding medium (see Chapter 4) or by taking advantage of *cathode protection* to prevent the nonsacrificial alloy component from dissolving, as will be shown in Chapter 5. These elements can also be synthesized using other alternative dealloying strategies, which will be discussed later. Elements highlighted in red in Fig. 3.5 cannot be synthesized in aqueous solutions due to their high reactivity. The second criterion is the reactivity with air. Nanoporous materials made of the elements in red in Fig. 3.5 are susceptible to spontaneously reacting with oxygen from air to form an oxide, usually through the following overall reaction [73].

$$\mathrm{nM\ (s)} + \frac{\mathrm{m}}{2}\ \mathrm{O_2\ (g)} \rightarrow \mathrm{M_nO_m\ (s)}\ \ (\Delta\mathrm{G} < 0) \tag{3.1}$$

In Eq. (3.1), M represents an element from the table in Fig. 3.2, and n and m represent the stoichiometry in the corresponding oxide [70]. The third criterion corresponds to their mechanical robustness for practical use. While some alkali metals like Li and Na are highly reactive with air and water, they are too ductile/soft to preserve their mechanical stability after nanostructuring and, therefore, are not included in Fig. 3.5. In the next section, we will provide a more detailed and comprehensive discussion on the chemical and electrochemical reaction mechanisms associated with the various dealloying methods developed in the community. We will discuss both precious nanoporous metals and air- and water-sensitive nanoporous metals.

3.4 Chemical and electrochemical reaction mechanisms in dealloying

We will start by elucidating the chemical reaction mechanisms associated with the fabrication of precious nanoporous metals by free corrosion dealloying and by electrolytic dealloying in aqueous solutions. First, we will use nanoporous gold (NP-Au) made from Au-Ag alloys as our model system to introduce the readers to the dealloying chemistries in general, which will later help understand the fundamental barriers encountered during the fabrication of air- and water-sensitive nanoporous metals and metalloids. Following this, we will discuss the reaction mechanisms associated with the fabrication of other nanoporous metals, including air- and water-sensitive nanoporous metals.

3.4.1 Free corrosion dealloying in aqueous solutions

When an Au-Ag alloy is exposed to an appropriate corroding medium, the underlying elements, Au and Ag, are susceptible to dissolving through the following oxidation reactions, respectively [38]:

$$Au\ (s) \rightarrow Au^{3+}(aq) + 3e^{-}(E_O = +1.50\ V\ vs.\ SHE) \tag{3.2}$$

$$Ag\ (s) \rightarrow Ag^{+}(aq) + e^{-}(E_O = +0.80\ V\ vs.\ SHE) \tag{3.3}$$

A voltage higher than the equilibrium potential of Au^{3+}/Au ($E_O = +1.50$ V vs. SHE) is required to dissolve Au through the oxidation reaction in Eq. (3.2). Similarly, a voltage higher than the equilibrium potential of Ag^{+}/Ag ($E_O = +0.80$ vs. SHE) is required to dissolve Ag through the oxidation reaction in Eq. (3.3). Therefore, each of these two oxidation reactions can only proceed if it is coupled with an appropriate reduction counter-reaction that will provide the required voltage and consume electrons released during metal oxidation to satisfy charge conservation [38]. In the case of free corrosion dealloying, the reduction counter-reaction involves the ionic species in the corroding medium with the highest standard reduction potential. To further illustrate this, let us examine the free corrosion behavior of an Au-Ag parent alloy in two different types of acids, namely a nonoxidizing acid and an oxidizing acid.

Case #1: *Non-oxidizing acid.* If a nonoxidizing acid such as 1 M of hydrochloric acid (HCl) is used for free corrosion dealloying of an Au-Ag parent alloy, the hydrogen evolution reaction (HER) given by Eq. (3.4) will be the reduction reaction with the highest standard reduction potential (assuming there is no oxygen in the electrolyte for the oxygen evolution reaction, coined ORR) [38].

$$2H^{+} + 2e^{-} \rightarrow H_2(g)(E_R = 0.0 \text{ V vs. SHE}) \tag{3.4}$$

Thermodynamically, the overall redox reaction for the free corrosion dealloying can only spontaneously proceed if its change in the Gibbs free energy is negative ($\Delta G<0$). Since ΔG is related to both the oxidation and reduction reaction potentials through the reaction in Eq. (3.5), where q represents the overall charge transferred during the process, ΔG will only be negative if E_R-E_O is positive.

$$\Delta G = -q(E_R - E_O) \tag{3.5}$$

If Au were to corrode, it would do so through the oxidation reaction in Eq. (3.2) in combination with the HER in Eq. (3.4). Thus, for Au corrosion, E_R-$E_O = -1.50$ V, which is negative. Similarly, Ag can only corrode through the oxidation reaction in Eq. (3.3) in combination with the HER in Eq. (3.4). Thus, for Ag corrosion, E_R-$E_O = -0.80$ V, which is also negative. Hence, neither Au nor Ag can dissolve in an HCl solution, meaning that NP-Au cannot be created by free corrosion dealloying in a nonoxidizing acid. These oxidation and reduction reactions are further illustrated graphically using Evans diagrams in Fig. 3.6a, where the activation polarization lines associated with Ag oxidation (solid red line) and hydrogen

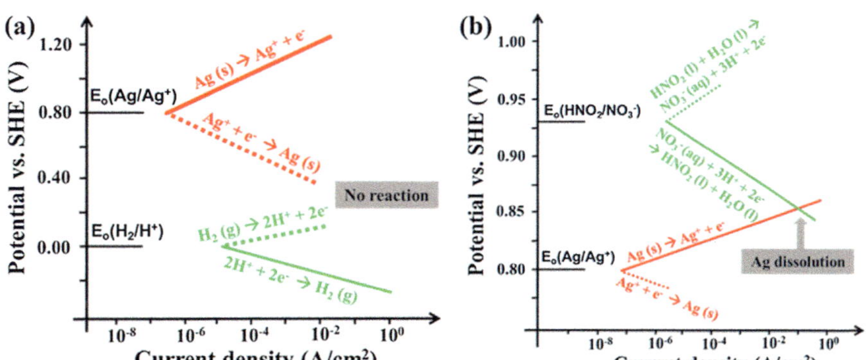

Figure 3.6 Use of Evans diagrams to explain the dissolution mechanisms of Ag in different corroding media. (a) In nonoxidizing acids, the cathodic reaction is proton reduction (solid green line), which does not intersect with Ag oxidization (solid red line), and thus Ag will not corrode. (b) In concentrated nitric acid, nitrate ion reduction (solid green line) intersects with Ag oxidation (solid red line), and Ag will corrode spontaneously. *Reproduced with permission from Ref. [38].*

reduction (solid green line) do not intersect, which means that the corresponding redox reaction cannot proceed.

Case #2: *Oxidizing acid.* If an oxidizing acid such as nitric acid is used, the reduction reaction with the highest standard reduction potential will proceed. For concentrated nitric acid ($\approx$15 M HNO_3), this corresponds to the reduction of nitrate ions (NO_3^-) given by Eq. (3.6) [37,74]:

$$NO_3^-(aq) + 3H^+(aq) + 2e^- \rightarrow HNO_2(l) + H_2O(l)(E_R = +0.93 \text{ V vs. SHE}) \quad (3.6)$$

In the case of Au oxidation through Eq. (3.2) in combination with the reduction counter-reaction in Eq. (3.6), E_R-E_O = −0.57 V, which is negative, meaning that Au cannot dissolve spontaneously in concentrated HNO_3. In the case of Ag oxidation through Eq. (3.3) in combination with the reduction counter-reaction in Eq. (3.6), E_R-E_O = +0.13 V, which is positive, meaning that Ag can dissolve spontaneously in concentrated HNO_3. Therefore, the maximum polarization available to drive the dissolution of Ag is 130 mV. Note that in practice, the effective polarization will be much lower due to reaction overpotentials (activation, charge transfer, and concentration overpotentials) [38]. These oxidation and reduction reactions are also illustrated graphically using Evans diagrams in Fig. 3.6b, where the activation polarization lines associated with Ag oxidation (solid red line) and NO_3^- reduction (solid green line) intersect, indicating that the overall redox reaction will take place spontaneously.

3.4.2 Electrolytic dealloying in aqueous solutions

As discussed earlier, NP-Au cannot be created by free corrosion dealloying of an Au-Ag parent alloy in a nonoxidizing acid (Fig. 3.6a) because Ag dissolution in a nonoxidizing acid is not thermodynamically favorable. However, NP-Au can still be created in a nonoxidizing acid (and even in neutral pH solutions [75]) using electrolytic dealloying. During electrolytic dealloying, external electrical energy in the form of an electrical voltage or current is used to drive the dissolution of Ag in a three-electrode electrochemical cell. A typical three-electrode electrochemical cell configuration consists of an Au-Ag parent alloy used as the working electrode (WE), a counter electrode (CE) commonly made of Pt, and a reference electrode (RE), e.g., Ag/AgCl. During selective electrolytic leaching, a constant potential (potentiostatic dealloying) is applied between the Au-Ag WE and the RE to drive the oxidation of Ag through Eq. (3.3). Alternatively, a constant current (galvanostatic dealloying) can be applied between the Au-Ag WE and

CE. Regarding the reduction counter-reaction, the oxidation reaction in Eq. (3.3) can be coupled with one of the following reduction reactions at the CE: the HER (Eq. 3.4), Ag deposition [75], or ORR, depending on the nature of the electrolyte, as recently reported by Welborn et al. [76] This process results in the formation of NP-Au. The theoretical minimum required voltage can be obtained from the potential difference between the reactions Eqs. (3.3) and (3.4), namely 0.80 V, as illustrated in Fig. 3.7. In real situations, however, one should account for reaction overpotentials. As such, potentials higher than this minimum are used.

3.4.3 Air-free electrolytic dealloying in anhydrous organic electrolytes

The air-free electrolytic dealloying approach is inspired by the rocking-chair battery technology, wherein the electrolyte typically consists of a working ion (e.g., Li^+ for Li-ion batteries) dissolved in an anhydrous organic solvent [77]. The general electrochemical cell configuration for air-free electrolytic dealloying consists of a three-electrode electrochemical cell with the parent alloy in the form M_xS_{1-x} (M represents the base material and S the sacrificial metal) used as the WE, the sacrificial metal S used as the CE and RE, and an anhydrous organic solvent containing S^{z+} cations of the sacrificial element used as an S^{z+} -conducting electrolyte. A simplified cell configuration is schematized in Fig. 3.8 together with the corresponding oxidation and reaction half-reactions taking place at the WE and CE, respectively:

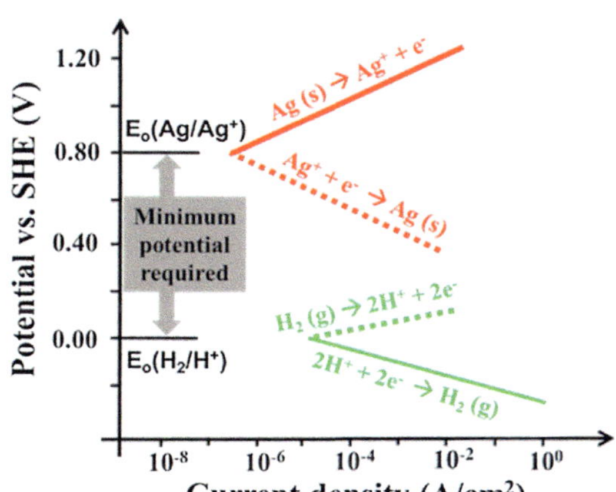

Figure 3.7 Use of Evans diagrams to elucidate the dissolution mechanisms of Ag during electrolytic dealloying, where external electrical energy supply is required.

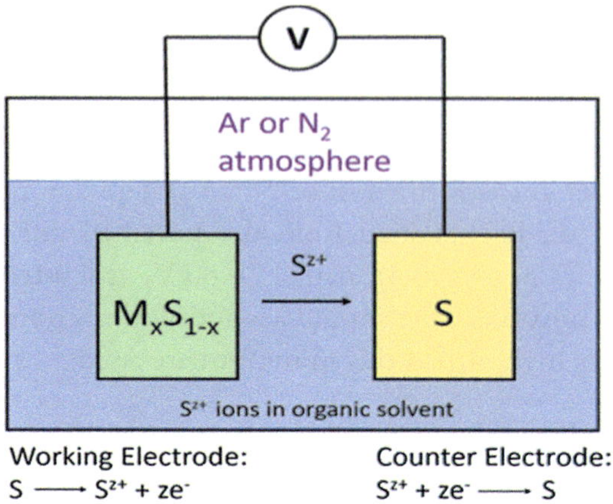

Figure 3.8 Schematic showing a generalized air-free electrolytic dealloying process, together with half-reactions on the WE and CE. Sacrificial element S is extracted from the parent alloy M_xS_{1-x} and plated onto the CE through the external application of a voltage or current.

$$\text{Working electrode}: \text{S} \rightarrow \text{S}^{z+} + \text{ze}^- \tag{3.7}$$

$$\text{Counter electrode}: \text{S}^{z+} + \text{ze}^- \rightarrow \text{S} \tag{3.8}$$

During electrolytic dealloying, the sacrificial metal S is removed from the WE through the oxidation reaction in Eq. (3.7) and plated on the CE through the reduction reaction in Eq. (3.8), either by applying a constant voltage between the WE and RE (i.e., potentiostatic mode) or by applying a constant current between the WE and CE (i.e., galvanostatic mode) [15,70]. Since electrolyte species are not consumed like with dealloying in aqueous solution, the overall air-free electrolytic dealloying concept is thus fundamentally different from conventional electrolytic dealloying in aqueous solutions, where species in the electrolyte are irreversibly reduced at the CE through HER or ORR as described above in Section 3.4.2. In addition, the use of an anhydrous organic solvent electrolyte under an inert environment (i.e., air-free electrolytic dealloying) makes it possible to fabricate air- and water-sensitive nanoporous materials [15,70]. Finally, the air-free electrolytic dealloying process makes it possible to recover the sacrificial element on the CE through the reduction counter-reaction in Eq. (3.8).

3.4.4 Vacuum thermal dealloying

Vacuum thermal dealloying is based on the difference in the partial vapor pressure between the components of a parent alloy. Upon heating a parent alloy under vacuum, the sacrificial component with the lower boiling point and higher partial vapor pressure (e.g., Mg or Zn) is selectively removed by sublimation (setup is shown in Fig. 3.9) [68]. The parent alloy, which is initially kept at a high temperature in a sealed furnace under inert gases (e.g., argon), is exposed to vacuum in order to remove the sacrificial element through sublimation [68,78]. The condensation unit at the end enables the recovery of the sublimated sacrificial element [68].

3.4.5 pH-controlled dealloying

pH-controlled dealloying is a form of free corrosion dealloying in which the pH of the corroding medium is adjusted to minimize the dissolution of the nonsacrificial component of an alloy. Typically, if the sacrificial and nonsacrificial components of a parent alloy are both made of nonprecious elements, both components can react with the aqueous corroding solution at different rates. Depending on how the pH is chosen, the dissolution of the sacrificial component can occur much faster than the dissolution of the nonsacrificial component. In selecting a pH value of the corroding medium, one can use Pourbaix diagrams or a simple potential-pH diagram like the one shown in Fig. 3.10, where the pH-dependent potential of the HER reduction counter-reaction involving H^+/H_2 from Eq. (3.4) (for an acidic corroding medium) or H_2O/H_2 from Eq. (3.9) (for an alkaline corroding medium) can be adjusted using the relation in Eq. (3.10) [80]:

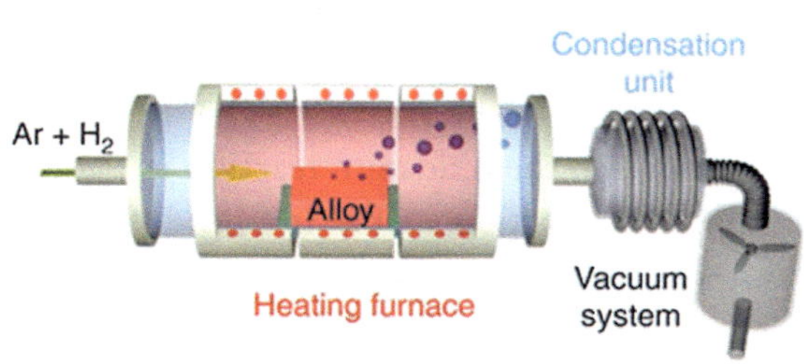

Figure 3.9 Schematic of a vacuum thermal dealloying setup. The alloy in the heating furnace is kept under vacuum, allowing for sublimation of the sacrificial element, which can be recovered at the condensation unit. *Reproduced with permission from Ref. [68].*

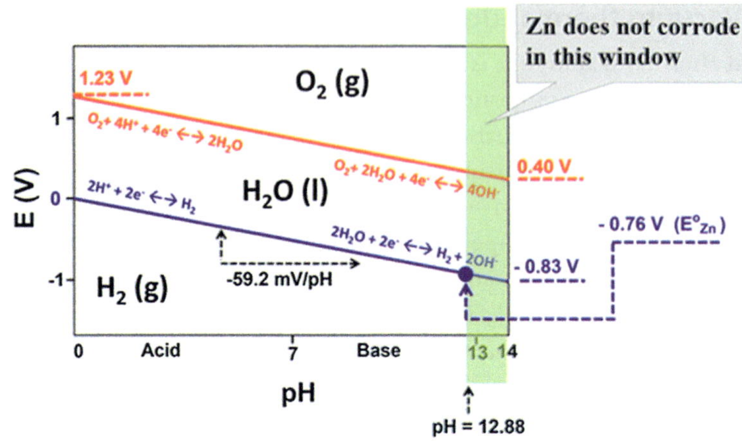

Figure 3.10 Potential–pH diagram showing the pH-dependent reduction potential of H+//H_2 (alkaline) reaction. Here, Zn will not corrode in the window highlighted in green. *Reproduced with permission from Ref. [79].*

$$2H_2O(l) + 2e^- \rightarrow H_2(g) + 2OH^-(aq)(\text{see Eq. (3.10)for } E_R) \tag{3.9}$$

$$E_R = -59.2 \text{ mV} \times \text{pH} \tag{3.10}$$

Let's use the formation of nanoporous Zn (NP-Zn) from a Zn-Al parent alloy as a model system to better illustrate the principle of pH-controlled dealloying. In this system, Al, the sacrificial component, can be selectively removed from the Zn-Al parent alloy by free corrosion in an aqueous corroding medium to form NP-Zn. At first glance, fabricating NP-Zn from Zn-Al alloy using an aqueous corroding medium may seem impossible because the standard reduction potentials of Zn and Al are -0.76 V and -1.66 V versus SHE, respectively, meaning that both Zn and Al should dissolve. If an alkaline solution with, e.g., pH = 12 is used as the corroding medium, the counter-reaction will be the HER in Eq. (3.9), and the corresponding reduction potential will be $E_R = -0.71$ V from Eq. (3.10). At this pH, Zn can dissolve through the oxidation reaction in Eq. (3.11) in combination with the reduction reaction in Eq. (3.9), since there is still a small driving force of E_R-E_O = +50 mV for Zn dissolution [79].

$$Zn(s) \rightarrow Zn^{2+}(aq) + 2e^-(E_O = -0.76 \text{ V vs. SHE}) \tag{3.11}$$

However, if a pH of 14 is used instead of 12, the potential for the HER deduced from Eq. (3.10) will become $E_R = -0.83$ V, and the driving force for Zn dissolution will become negative (i.e., E_R-E_O = -70 mV), meaning that at this pH, Zn should not dissolve through Eq. (3.11). On the other hand, at pH = 14, Al can dissolve through the oxidation reaction in Eq. (3.12) in combination with the reduction reaction in Eq. (3.9), since there is a large driving force of E_R-E_O = +830 mV available for Al dissolution [79].

$$Al(s) \rightarrow Al^{3+}(aq) + 3e^{-} (E_O = -1.66 \text{ V vs. SHE}) \quad (3.12)$$

Thus, there is a pH window between 12.9 and 14.0 where Al can selectively dissolve without attacking Zn, as highlighted in green in Fig. 3.7, and this pH window is used to fabricate NP-Zn [79]. In practice, once Al is fully dissolved, if the NP-Zn is kept in the corroding medium for a longer time, it will also dissolve, which suggests that in reality, both Zn and Al can dissolve in the pH window between 12.9 and 14. However, the dissolution rate of Al is significantly faster than that of Zn.

3.4.6 Dealloying by galvanic replacement

Dealloying by galvanic replacement is another form of free corrosion dealloying in which the reduction of a metal ion, instead of a non-metal ion, drives the corrosion process. In general, two half-reactions take place during galvanic replacement: the oxidation of a metal with a relatively lower reduction potential and the reduction of a metal ion with a relatively higher reduction potential. To illustrate the general principle, we use a Cu rod immersed in a solution containing silver ions (Ag^{+}) from an $AgNO_3$ solution, as shown in Fig. 3.11 [81]. Since Cu^{2+}/Cu (+0.34 V vs. SHE) has a lower standard reduction potential compared to Ag^{+}/Ag (+0.80 V vs. SHE), Cu atoms on the surface of the rod will dissolve into the solution through the oxidation reaction in Eq. (3.13) (Fig. 3.11a) [81]:

$$Cu(s) \rightarrow Cu^{2+}(aq) + 2e^{-} (E_O = +0.34 \text{ V vs. SHE}) \quad (3.13)$$

The Cu oxidation reaction in Eq. (3.13) is coupled with the reduction of Ag^{+} to metallic Ag through Eq. (3.14); see Fig. 3.11a [81]:

$$Ag^{+}(aq) + e^{-} \rightarrow Ag(s) (E_R = +0.80 \text{ V vs. SHE}) \quad (3.14)$$

The overall driving force for the galvanic replacement reaction is therefore given by E_R-E_O = +0.46 V, which is positive, meaning that Cu will spontaneously dissolve. In this specific example, the initially colorless $AgNO_3$ solution turns blue when Cu^{2+} ions are released into the solution (Fig. 3.11b) [81].

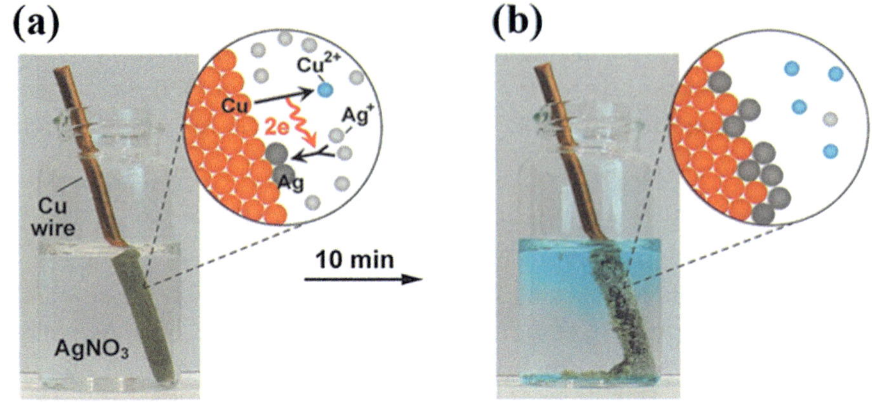

Figure 3.11 Illustration of a galvanic replacement mechanism using a Cu rod immersed in an aqueous solution containing Ag^+ from $AgNO_3$. (a) Cu rod directly after immersion; and (b) after the reaction had proceeded for 10 minutes. Cu dissolves in the solution (blue color) while metallic Ag is deposited onto the rod. The inset schematics show the corresponding atomic models. *Reproduced with permission from Ref. [81].*

3.4.7 Dealloying by reduction-induced decomposition

Reduction-induced decomposition is based on the selective dissolution of an ionic or covalent compound instead of selectively dissolving an alloy component [82—84]. For example, nanoporous silver (NP-Ag) can be made by chemically or electrolytically reducing an AgCl or Ag_2O_3 compound using a sacrificial reducing agent (in the case of chemical reduction) or a negative voltage or current (in the case of electrolytic reduction). During the chemical reduction of a particle made of AgCl or Ag_2O_3 compound, the reaction proceeds uniformly from the particle/electrolyte interface inwards, as illustrated in Fig. 3.12a [82]. Due to the poor electrical conductivity of the compound, the reaction is initially restricted to the substrate/compound/electrolyte triple phase boundaries [82,85]. The reduced metal, however, is conductive and serves as a new current collector, which moves the reaction front along the compound surface and eventually transforms the entire sample (Fig. 3.12b) [82]. In the case of the electrolytic reduction, the reduction half-reaction taking place at the WE requires an oxidation counter half-reaction to supply electrons, meaning that a sacrificial material is oxidized at the CE.

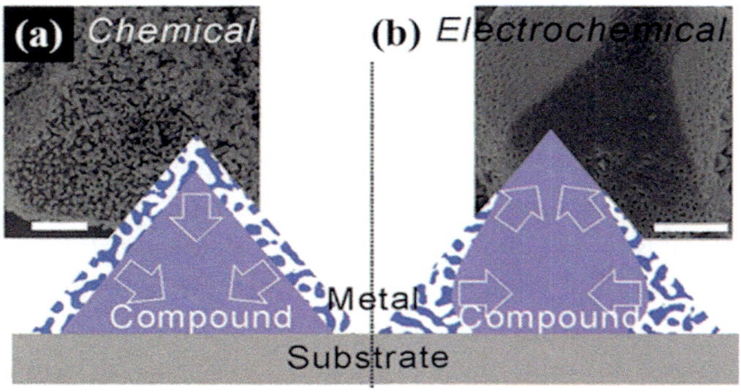

Figure 3.12 Schematic illustrations of chemically (a) and electrolytically (b) reduced Ag_2O_3 crystals. In chemical reduction (a), the reaction front propagates uniformly. In electrolytic reduction (b), the reaction front propagates from the bottom of the crystal where it is in contact with the conducting substrate. Scale bars are 2 μm. *Reproduced with permission from Ref. [82].*

3.4.8 Dealloying by thermal decomposition

During the fabrication of nanoporous metals by thermal decomposition, thermal reduction is used to selectively remove the sacrificial component from a parent material, generally consisting of a transition metal dichalcogenide or a metal salt. A typical setup used for dealloying by thermal decomposition is shown in Fig. 3.13 [86]. The thermal decomposition is carried out around 1000°C under an H_2 or Ar atmosphere [86]. The removal of the sacrificial element occurs through two parallel processes, both of which are associated with breaking covalent bonds at the gas-solid interface [86]: (i) heat-induced evaporation of the high vapor pressure chalcogen elements

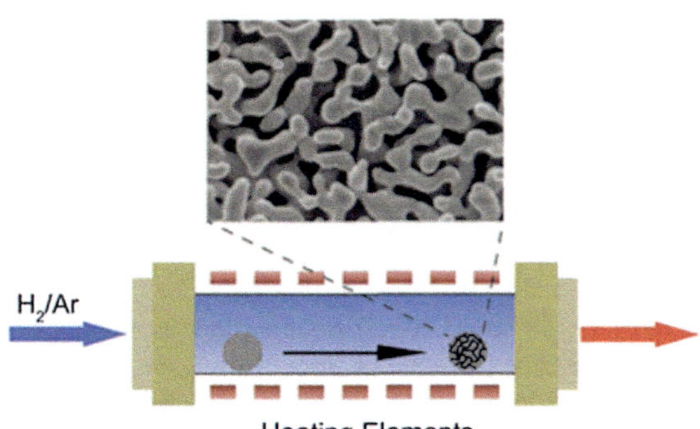

Figure 3.13 Schematic illustration of thermal decomposition procedure. *Reproduced from Ref. [86].*

[68,86] and (ii) reaction-induced removal through hydrogen gas as shown in Eq. (3.15) [86]:

$$MX\ (s) + H_2(g) \rightarrow M(s) + H_2X(g), X = S\ /\ Se \tag{3.15}$$

Two key differences between thermal decomposition and the vacuum thermal dealloying method described in section (d) are (i) the use of a nonmetal, such as a chalcogen (S, Se, or Te), as the sacrificial element in thermal decomposition, while metals with low boiling points and high vapor pressures are used as sacrificial elements during vacuum thermal dealloying; (ii) the use of atmospheric pressure and a reducing gas during thermal decomposition [86,87], versus the use of vacuum during vacuum thermal dealloying.

3.4.9 Liquid metal dealloying

Liquid metal dealloying is based on the selective dissolution of the sacrificial components of an alloy using a metallic melt as the corroding medium. Central to this dealloying approach is the interplay between the enthalpies of mixing (ΔH_{mix}) of the parent alloy components and the metallic melt: If the sacrificial component in the parent alloy has a negative enthalpy of mixing with the metallic melt element, the melt will dissolve the sacrificial component. Thus, if the nonsacrificial component in the parent alloy has a positive enthalpy of mixing with the metallic melt element, it will not be dissolved [88]. In other words, the metallic melt dissolves away the component in the parent alloy that it can more easily mix with and leaves behind the component it cannot mix with. This is schematically illustrated in Fig. 3.14a and b in the case of a Mg-Si parent alloy system in a Bi melt

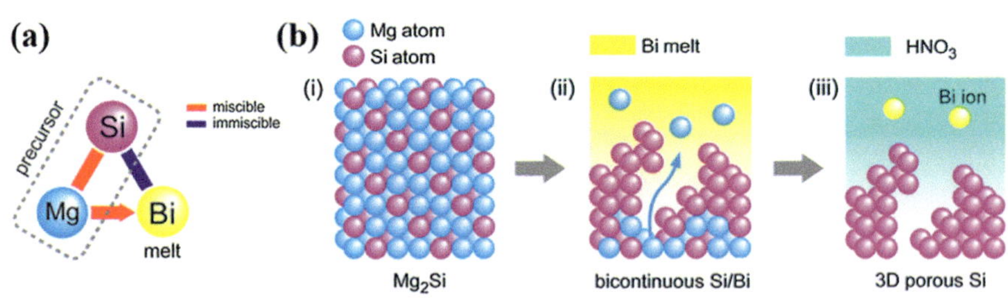

Figure 3.14 (a) Schematic of liquid metal dealloying's working principle using the Si-Mg-Bi system. Here, the Si-Mg precursor and Bi melt were selected based on the miscibility of Mg-Bi and the immiscibility of Si-Bi. (b) Schematic of the evolution of the nanoporous structure during liquid metal dealloying. The solidified Bi melt requires an additional etching step in HNO_3, as shown in (iii). *Reproduced with permission from Ref. [89].*

[67,89]. Note that a postprocessing step is further required to remove the solidified metallic melt. This is commonly done by using an aqueous corroding medium to dissolve the metallic melt, as illustrated in Fig. 3.14b. It should also be emphasized that since during liquid metal dealloying the dissolution of the sacrificial component is driven by the enthalpy of mixing, the sacrificial component can be more noble than the nonsacrificial component. To illustrate, even though Cu is more noble than Ti, nanoporous Ti can be made by selective removal of Cu from Ti-Cu parent alloys using Mg melt. Cu and Mg melt readily mix, in contrast to Ti and Cu. So, Cu will dissolve in the Mg melt to enable the formation of nanoporous Ti [90].

3.5 Conclusions

In this chapter, we have elucidated the fundamental mechanism that governs the formation of nanoporosity via a spinodal decomposition pathway during dealloying. We have also elucidated the chemical and electrochemical reaction mechanisms associated with state-of-the-art dealloying methods developed by the community. Finally, we have presented the common dealloying routes reported in the literature for the fabrication of non-precious nanoporous metals: Air-free electrolytic dealloying in anhydrous organic solvent electrolytes is very suitable for the production of the chemically reactive nanoporous elements highlighted in red in Fig. 3.5. The main advantages of this method are the possibility of producing nearly oxide-free nanoporous metals and the recovery of the sacrificial component at the counter electrode. The vacuum thermal dealloying, which can also be used to fabricate nearly oxide-free nanoporous materials from the reactive elements highlighted in red and green in Fig. 3.5. The postreaction condenser enables the recovery of sacrificial elements. pH-controlled dealloying and dealloying by galvanic replacement. Dealloying by reduction-induced decomposition and thermal decomposition, in which the parent material is a compound precursor instead of an alloy-type precursor. Liquid metal dealloying, which was one of the first methods that enabled the fabrication of less-noble nanoporous materials.

Acknowledgments

Part of the content of this chapter has been published by J. Fu, S.S. Welborn, and E. Detsi. Invited Contribution for ACS Applied Energy Materials 5 (2022) entitled "Dealloyed Air- and Water-Sensitive Nanoporous Metals and Metalloids for Emerging Energy Applications."

The authors thank Jintao Fu, John Corsi, Sam Welborn, and Alexander Ng for their discussion and contributions.
Another part of the content of this chapter was obtained from John S. Corsi's dissertation under the supervision of E. Detsi entitled *Fabrication and Study of Microstructure, Solid-Electrolyte Interphase, and Phase Evolution of Nanoporous Metals Used as Li-Ion Battery Alloy Anodes* (UPenn, 2021). https://repository.upenn.edu/entities/publication/39e8d876-f749-4140-916b-6728148bcf63. The authors thank John Corsi for the discussion and contributions.

References

[1] J. Fu, S.S. Welborn, E. Detsi, Dealloyed air- and water-sensitive nanoporous metals and metalloids for emerging energy applications, ACS Appl. Energy Mater. 5 (2022) 6516–6544, https://doi.org/10.1021/acsaem.2c00405.
[2] Y. Ding, Y.-J. Kim, J. Erlebacher, Nanoporous gold leaf: "Ancient technology"/advanced material, Adv. Mater. 16 (2004) 1897–1900, https://doi.org/10.1002/adma.200400792.
[3] Q. Sang, S. Hao, J. Han, Y. Ding, Dealloyed nanoporous materials for electrochemical energy conversion and storage, EnergyChem 4 (2022) 100069, https://doi.org/10.1016/j.enchem.2022.100069.
[4] J. Fu, E. Detsi, J. Th. M. De Hosson, Recent advances in nanoporous materials for renewable energy resources conversion into fuels, Surf. Coat. Technol. 347 (2018) 320–336, https://doi.org/10.1016/j.surfcoat.2018.05.001.
[5] B. Zeifert, J.S. Blásquez, J.G.C. Moreno, H.A. Calderón, Raney-nickel catalysts produced by mechanical alloying, Rev. Adv. Mater. Sci. 18 (2008) 632–638.
[6] T.W.G. Solomons, C.B. Fryhle, Organic Chemistry, 2012, https://doi.org/10.1007/s13398-014-0173-7.2.
[7] A. Wittstock, V. Zielasek, J. Biener, C.M. Friend, M. Bäumer, Nanoporous gold catalysts for selective gas-phase oxidative coupling of methanol at low temperature, Science 327 (2010) 319–322, https://doi.org/10.1126/science.1183591.
[8] V. Zielasek, B. Jürgens, C. Schulz, J. Biener, M.M. Biener, A. V Hamza, M. Bäumer, Gold catalysts: nanoporous gold foams, Angew. Chem. Int. Ed. 45 (2006) 8241–8244, https://doi.org/10.1002/anie.200602484.
[9] M.M. Biener, J. Biener, A. Wichmann, A. Wittstock, T.F. Baumann, M. Bäumer, A.V. Hamza, ALD functionalized nanoporous gold: thermal stability, mechanical properties, and catalytic activity, Nano Lett. 11 (2011) 3085–3090, https://doi.org/10.1021/nl200993g.
[10] Q. Lu, J. Rosen, Y. Zhou, G.S. Hutchings, Y.C. Kimmel, J.G. Chen, F. Jiao, A selective and efficient electrocatalyst for carbon dioxide reduction, Nat. Commun. 5 (2014) 3242, https://doi.org/10.1038/ncomms4242.
[11] J. Zhang, C.M. Li, Nanoporous metals: fabrication strategies and advanced electrochemical applications in catalysis, sensing and energy systems, Chem. Soc. Rev. 41 (2012) 7016–7031, https://doi.org/10.1039/C2CS35210A.
[12] L. Gan, M. Heggen, R. O'Malley, B. Theobald, P. Strasser, Understanding and controlling nanoporosity formation for improving the stability of bimetallic fuel cell catalysts, Nano Lett. 13 (2013) 1131–1138, https://doi.org/10.1021/nl304488q.
[13] W. Luc, F. Jiao, Nanoporous metals as electrocatalysts: state-of-the-art, opportunities, and challenges, ACS Catal. 7 (2017) 5856–5861, https://doi.org/10.1021/acscatal.7b01803.
[14] J.B. Cook, T.C. Lin, E. Detsi, J. Nelson Weker, S.H. Tolbert, J.B. Cook, T.C. Lin, E. Detsi, J.N. Weker, S.H. Tolbert, Using X-ray microscopy to understand how nanoporous materials can be used to reduce the large volume change in alloy Anodes, Nano Lett. 17 (2017) 870–877, https://doi.org/10.1021/acs.nanolett.6b04181.

[15] J.S. Corsi, J. Fu, Z. Wang, T. Lee, A.K. Ng, E. Detsi, Hierarchical bulk nanoporous aluminum for on-site generation of hydrogen by hydrolysis in pure water and combustion of solid fuels, ACS Sustain. Chem. Eng. 7 (2019) 11194–11204, https://doi.org/10.1021/acssuschemeng.9b00481.
[16] Y. Yang, G. Ruan, C. Xiang, G. Wang, J.M. Tour, Flexible three-dimensional nanoporous metal-based energy devices, J. Am. Chem. Soc. 136 (2014) 6187–6190, https://doi.org/10.1021/ja501247f.
[17] X. Lang, A. Hirata, T. Fujita, M. Chen, Nanoporous metal/oxide hybrid electrodes for electrochemical supercapacitors, Nat. Nanotechnol. 6 (2011) 232–236, https://doi.org/10.1038/nnano.2011.13.
[18] S. Kondrat, C.R. Pérez, V. Presser, Y. Gogotsi, A.A. Kornyshev, Effect of pore size and its dispersity on the energy storage in nanoporous supercapacitors, Energy Environ. Sci. 5 (2012) 6474–6479, https://doi.org/10.1039/C2EE03092F.
[19] J. Han, Y.-C. Lin, L. Chen, Y.-C. Tsai, Y. Ito, X. Guo, A. Hirata, T. Fujita, M. Esashi, T. Gessner, M. Chen, On-chip micro-pseudocapacitors for ultrahigh energy and power delivery, Adv. Sci. 2 (2015) 1500067, https://doi.org/10.1002/advs.201500067.
[20] Y.-G. Guo, J.-S. Hu, L.-J. Wan, Nanostructured materials for electrochemical energy conversion and storage devices, Adv. Mater. 20 (2008) 2878–2887, https://doi.org/10.1002/adma.200800627.
[21] Y. An, Y. Tian, C. Wei, Y. Tao, B. Xi, S. Xiong, J. Feng, Y. Qian, Dealloying: an effective method for scalable fabrication of 0D, 1D, 2D, 3D materials and its application in energy storage, Nano Today 37 (2021) 101094, https://doi.org/10.1016/j.nantod.2021.101094.
[22] X. Wu, G. He, Y. Ding, Dealloyed nanoporous materials for rechargeable post-lithium batteries, ChemSusChem 13 (2020) 3376–3390, https://doi.org/10.1002/cssc.202001369.
[23] E. Detsi, P. Onck, J. Th, M. De Hosson, J. Th, M. De Hosson, Metallic muscles at work: high rate actuation in nanoporous gold/polyaniline composites, ACS Nano 7 (2013) 4299–4306, https://doi.org/10.1021/nn400803x.
[24] J. Biener, A. Wittstock, L.A. Zepeda-Ruiz, M.M. Biener, V. Zielasek, D. Kramer, R.N. Viswanath, J. Weissmüller, M. Bäumer, A. V Hamza, Surface-chemistry-driven actuation in nanoporous gold, Nat. Mater. 8 (2009) 47–51, https://doi.org/10.1038/nmat2335.
[25] H.J. Jin, X.L. Wang, S. Parida, K. Wang, M. Seo, J. Weissmüller, Nanoporous Au-Pt alloys as large strain electrochemical actuators, Nano Lett. 10 (2010) 187–194, https://doi.org/10.1021/nl903262b.
[26] H.-J. Jin, J. Weissmüller, Bulk nanoporous metal for actuation, Adv. Eng. Mater. 12 (2010) 714–723, https://doi.org/10.1002/adem.200900329.
[27] A. Wittstock, J. Biener, M. Bäumer, Nanoporous gold: a new material for catalytic and sensor applications, Phys. Chem. Chem. Phys. 12 (2010) 12919–12930, https://doi.org/10.1039/C0CP00757A.
[28] L.Y. Chen, X.Y. Lang, T. Fujita, M.W. Chen, Nanoporous gold for enzyme-free electrochemical glucose sensors, Scripta Mater. 65 (2011) 17–20, https://doi.org/10.1016/j.scriptamat.2011.03.025.
[29] Z. Liu, P.C. Searson, Single nanoporous gold nanowire sensors, J. Phys. Chem. B 110 (2006) 4318–4322, https://doi.org/10.1021/jp056940t.
[30] L.-Y.Y. Chen, J.-S.S. Yu, T. Fujita, M.-W.W. Chen, Nanoporous copper with tunable nanoporosity for SERS applications, Adv. Funct. Mater. 19 (2009) 1221–1226, https://doi.org/10.1002/adfm.200801239.
[31] L.H. Qian, X.Q. Yan, T. Fujita, A. Inoue, M.W. Chen, Surface enhanced Raman scattering of nanoporous gold: smaller pore sizes stronger enhancements, Appl. Phys. Lett. 90 (2007) 153120, https://doi.org/10.1063/1.2722199.

[32] S.O. Kucheyev, J.R. Hayes, J. Biener, T. Huser, C.E. Talley, A. V Hamza, Surface-enhanced Raman scattering on nanoporous Au, Appl. Phys. Lett. 89 (2006) 53102, https://doi.org/10.1063/1.2260828.

[33] J. Biener, G.W. Nyce, A.M. Hodge, M.M. Biener, A. V Hamza, S.A. Maier, Nanoporous plasmonic metamaterials, Adv. Mater. 20 (2008) 1211–1217, https://doi.org/10.1002/adma.200701899.

[34] A. Pröschel, J. Chacko, R. Whitaker, M.A.U. Chen, E. Detsi, Visible light plasmonic heating-enhanced electrochemical current in nanoporous gold cathodes, J. Electrochem. Soc. 166 (2019) H146–H150, https://doi.org/10.1149/2.0601904jes.

[35] S. Ahl, P.J. Cameron, J. Liu, W. Knoll, J. Erlebacher, F. Yu, A comparative plasmonic study of nanoporous and evaporated gold films, Plasmonics 3 (2008) 13–20, https://doi.org/10.1007/s11468-007-9048-5.

[36] J. Erlebacher, M.J. Aziz, A. Karma, N. Dimitrov, K. Sieradzki, Evolution of nanoporosity in dealloying, Nature 410 (2001) 450–453, https://doi.org/10.1038/35068529.

[37] J. Fu, J.S. Corsi, Z. Wang, H. Wei, E. Detsi, Integrated metal-air battery and selective electrolytic leaching cell for the preparation of nanoporous metals, ACS Appl. Nano Mater. 1 (2018) 4164–4169, https://doi.org/10.1021/acsanm.8b00919.

[38] J. Fu, Z. Deng, E. Detsi, Nanoporous gold formation by free corrosion dealloying of gold-silver alloys in nonoxidizing acids driven by catalytic oxygen reduction reaction, Journal of Metals 71 (2019) 1581–1589, https://doi.org/10.1007/s11837-019-03383-1.

[39] E. Detsi, S. Punzhin, J. Rao, P.R. Onck, J. Th, M. De Hosson, Enhanced strain in functional nanoporous gold with a dual microscopic length scale structure, ACS Nano 6 (2012) 3734–3744, https://doi.org/10.1021/nn300179n.

[40] E. Detsi, M. Salverda, P.R. Onck, J. Th, M. De Hosson, On the localized surface plasmon resonance modes in nanoporous gold films, J. Appl. Phys. 115 (2014) 044308, https://doi.org/10.1063/1.4862440.

[41] Z. Deng, E. Detsi, Enhancing the free corrosion dealloying rate with a catalytically driven reaction, Nanoscale 9 (2017) 11858–11863, https://doi.org/10.1039/C7NR04611A.

[42] H.-J. Jin, D. Kramer, Y. Ivanisenko, J. Weissmüller, Macroscopically strong nanoporous Pt prepared by dealloying, Adv. Eng. Mater. 9 (2007) 849–854, https://doi.org/10.1002/adem.200700177.

[43] D. V Pugh, A. Dursun, S.G. Corcoran, Formation of nanoporous platinum by selective dissolution of Cu from Cu0.75Pt0.25, J. Mater. Res. 18 (2003) 216–221, https://doi.org/10.1557/JMR.2003.0030.

[44] X. Wang, W. Wang, Z. Qi, C. Zhao, H. Ji, Z. Zhang, Electrochemical catalytic activities of nanoporous palladium rods for methanol electro-oxidation, J. Power Sources 195 (2010) 6740–6747, https://doi.org/10.1016/j.jpowsour.2010.03.098.

[45] X. Wang, W. Wang, Z. Qi, C. Zhao, H. Ji, Z. Zhang, High catalytic activity of ultrafine nanoporous palladium for electro-oxidation of methanol, ethanol, and formic acid, Electrochem. Commun. 11 (2009) 1896–1899, https://doi.org/10.1016/j.elecom.2009.08.011.

[46] J.R. Hayes, A.M. Hodge, J. Biener, A. V Hamza, K. Sieradzki, Monolithic nanoporous copper by dealloying Mn–Cu, J. Mater. Res. 21 (2006) 2611–2616, https://doi.org/10.1557/jmr.2006.0322.

[47] Y. Ding, M. Chen, Nanoporous metals for catalytic and optical applications, MRS Bull. 34 (2009) 569–576, https://doi.org/10.1557/mrs2009.156.

[48] Y. Bing, H. Liu, L. Zhang, D. Ghosh, J. Zhang, Nanostructured Pt-alloy electrocatalysts for PEM fuel cell oxygen reduction reaction, Chem. Soc. Rev. 39 (2010) 2184–2202, https://doi.org/10.1039/B912552C.

[49] F. Cheng, J. Chen, Metal—air batteries: from oxygen reduction electrochemistry to cathode catalysts, Chem. Soc. Rev. 41 (2012) 2172—2192, https://doi.org/10.1039/C1CS15228A.
[50] Y. Nie, L. Li, Z. Wei, Recent advancements in Pt and Pt-free catalysts for oxygen reduction reaction, Chem. Soc. Rev. 44 (2015) 2168—2201, https://doi.org/10.1039/C4CS00484A.
[51] D.S. Sundaram, V. Yang, V.E. Zarko, Combustion of nano aluminum particles (Review), Combust. Explos. Shock Waves 51 (2015) 173—196, https://doi.org/10.1134/S0010508215020045.
[52] Y. Huang, G.A. Risha, V. Yang, R.A. Yetter, Effect of particle size on combustion of aluminum particle dust in air, Combust. Flame 156 (2009) 5—13, https://doi.org/10.1016/j.combustflame.2008.07.018.
[53] H. Tyagi, P.E. Phelan, R. Prasher, R. Peck, T. Lee, J.R. Pacheco, P. Arentzen, Increased hot-plate ignition probability for nanoparticle-laden diesel fuel, Nano Lett. 8 (2008) 1410—1416, https://doi.org/10.1021/nl080277d.
[54] T. Miller, J. Herr, Green rocket propulsion by reaction of Al and Mg powders and water, in: 40th AIAA/ASME/SAE/ASEE Joint Propulsion Conference and Exhibit, American Institute of Aeronautics and Astronautics, 2004, https://doi.org/10.2514/6.2004-4037.
[55] A. Ingenito, C. Bruno, Using aluminum for space propulsion, J. Propul. Power 20 (2004) 1056—1063, https://doi.org/10.2514/1.5132.
[56] G. Risha, T. Connell, R. Yetter, V. Yang, T. Wood, M. Pfeil, T. Pourpoint, S. Son, Aluminum-ICE (ALICE) propellants for hydrogen generation and propulsion, in: 45th AIAA/ASME/SAE/ASEE Joint Propulsion Conference & Exhibit, American Institute of Aeronautics and Astronautics, 2009, https://doi.org/10.2514/6.2009-4877.
[57] J. Wang, D. Gao, G. Wang, S. Miao, H. Wu, J. Li, X. Bao, Cobalt nanoparticles encapsulated in nitrogen-doped carbon as a bifunctional catalyst for water electrolysis, J. Mater. Chem. A Mater. 2 (2014) 20067—20074, https://doi.org/10.1039/C4TA04337E.
[58] N.A.M. Barakat, M. Motlak, A.A. Elzatahry, K.A. Khalil, E.A.M. Abdelghani, NixCo1—x alloy nanoparticle-doped carbon nanofibers as effective non-precious catalyst for ethanol oxidation, Int. J. Hydrogen Energy 39 (2014) 305—316, https://doi.org/10.1016/j.ijhydene.2013.10.061.
[59] W. Sheng, A.P. Bivens, M. Myint, Z. Zhuang, R.V. Forest, Q. Fang, J.G. Chen, Y. Yan, Non-precious metal electrocatalysts with high activity for hydrogen oxidation reaction in alkaline electrolytes, Energy Environ. Sci. 7 (2014) 1719—1724, https://doi.org/10.1039/c3ee43899f.
[60] S. Litvinenko, S. Alekseev, V. Lysenko, A. Venturello, F. Geobaldo, L. Gulina, G. Kuznetsov, V. Tolstoy, V. Skryshevsky, E. Garrone, D. Barbier, Hydrogen production from nano-porous Si powder formed by stain etching, Int. J. Hydrogen Energy 35 (2010) 6773—6778, https://doi.org/10.1016/j.ijhydene.2010.04.041.
[61] H.Z. Wang, D.Y.C. Leung, M.K.H. Leung, M. Ni, A review on hydrogen production using aluminum and aluminum alloys, Renew. Sustain. Energy Rev. 13 (2009) 845—853, https://doi.org/10.1016/j.rser.2008.02.009.
[62] F. Erogbogbo, T. Lin, P.M. Tucciarone, K.M. LaJoie, L. Lai, G.D. Patki, P.N. Prasad, M.T. Swihart, On-demand hydrogen generation using nanosilicon: Splitting water without light, heat, or Electricity, Nano Lett. 13 (2013) 451—456, https://doi.org/10.1021/nl304680w.
[63] H. Yaghoobnejad Asl, J. Fu, H. Kumar, S.S. Welborn, V.B. Shenoy, E. Detsi, In situ dealloying of bulk Mg_2Sn in Mg-ion half cell as an effective route to nanostructured Sn for high performance Mg-ion battery anodes, Chem. Mater. 30 (2018) 1815—1824, https://doi.org/10.1021/acs.chemmater.7b04124.

[64] M. Li, T. Qiu, A.C. Foucher, J. Fu, Z. Wang, D. Zhang, A.M. Rappe, E.A. Stach, E. Detsi, Impact of hierarchical nanoporous architectures on sodium storage in antimony-based sodium-ion battery anodes, ACS Appl. Energy Mater. 3 (2020) 11231−11241, https://doi.org/10.1021/acsaem.0c02161.
[65] D. Zhang, J. Fu, Z. Wang, L. Wang, J.S. Corsi, E. Detsi, Perspective—reversible magnesium storage in silicon: an ongoing challenge, J. Electrochem. Soc. 167 (2020) 050514, https://doi.org/10.1149/1945-7111/ab736b.
[66] J.B. Cook, E. Detsi, Y. Liu, Y.-L. Liang, H.-S. Kim, X. Petrissans, B. Dunn, S.H. Tolbert, Nanoporous tin with a granular hierarchical ligament morphology as a highly stable Li-ion battery anode, ACS Appl. Mater. Interfaces 9 (2017) 293−303, https://doi.org/10.1021/acsami.6b09014.
[67] I. McCue, B. Gaskey, P.-A. Geslin, A. Karma, J. Erlebacher, Kinetics and morphological evolution of liquid metal dealloying, Acta Mater. 115 (2016) 10−23, https://doi.org/10.1016/j.actamat.2016.05.032.
[68] Z. Lu, C. Li, J. Han, F. Zhang, P. Liu, H. Wang, Z. Wang, C. Cheng, L. Chen, A. Hirata, T. Fujita, J. Erlebacher, M. Chen, Three-dimensional bicontinuous nanoporous materials by vapor phase dealloying, Nat. Commun. 9 (2018) 276, https://doi.org/10.1038/s41467-017-02167-y.
[69] Q. Chen, K. Sieradzki, Spontaneous evolution of bicontinuous nanostructures in dealloyed Li-based systems, Nat. Mater. 12 (2013) 1102−1106, https://doi.org/10.1038/nmat3741.
[70] J. Fu, J.S. Corsi, S.S. Welborn, V. Basile, L. Wang, A.K. Ng, E. Detsi, Eco-friendly synthesis of nanoporous magnesium by air-free electrolytic dealloying with recovery of sacrificial elements for energy conversion and storage applications, ACS Sustain. Chem. Eng. 9 (2021) 2762−2769, https://doi.org/10.1021/acssuschemeng.0c08157.
[71] P. Vanysek, Electrochemical series, in: Handbook of Chemistry and Physics, Hand Book, 2000.
[72] J. Shen, S. Liu, Z. He, L. Shi, Influence of antimony ions in negative electrolyte on the electrochemical performance of vanadium redox flow batteries, Electrochim. Acta 151 (2015) 297−305, https://doi.org/10.1016/J.ELECTACTA.2014.11.060.
[73] O. Palanna, Engineering Chemistry, 2017, https://doi.org/10.1017/cbo9781316146743.
[74] P. Fauvet, F. Balbaud, R. Robin, Q.-T. Tran, A. Mugnier, D. Espinoux, Corrosion mechanisms of austenitic stainless steels in nitric media used in reprocessing plants, J. Nucl. Mater. 375 (2008) 52−64, https://doi.org/10.1016/j.jnucmat.2007.10.017.
[75] J. Snyder, K. Livi, J. Erlebacher, Dealloying silver/gold alloys in neutral silver nitrate solution: porosity evolution, surface composition, and surface oxides, J. Electrochem. Soc. 155 (2008) C464−C473, https://doi.org/10.1149/1.2940319.
[76] S.S. Welborn, J.S. Corsi, L. Wang, A. Lee, J. Fu, E. Detsi, Effects of side reactions on the kinetics of nanoporous gold formation revealed by real-time X-ray scattering during electrolytic dealloying, J. Mater. Chem. A Mater. 9 (2021) 19994−20005, https://doi.org/10.1039/D1TA04822H.
[77] B. Scrosati, Lithium rocking chair batteries: an old concept? J. Electrochem. Soc. 139 (1992) 2776−2781, https://doi.org/10.1149/1.2068978.
[78] M. Kosmidou, M.J. Detisch, T.L. Maxwell, T.J. Balk, Vacuum thermal dealloying of magnesium-based alloys for fabrication of nanoporous refractory metals, MRS Commun. 9 (2019) 144−149, https://doi.org/10.1557/mrc.2019.15.
[79] J. Fu, Z. Deng, T. Lee, J.S. Corsi, Z. Wang, D. Zhang, E. Detsi, pH-controlled dealloying route to hierarchical bulk nanoporous Zn derived from metastable alloy for hydrogen generation by hydrolysis of Zn in neutral water, ACS Appl. Energy Mater. 1 (2018) 3198−3205, https://doi.org/10.1021/acsaem.8b00419.

[80] M.G. Walter, E.L. Warren, J.R. McKone, S.W. Boettcher, Q. Mi, E.A. Santori, N.S. Lewis, Solar water splitting cells, Chem. Rev. 110 (2010) 6446–6473, https://doi.org/10.1021/cr1002326.

[81] X. Xia, Y. Wang, A. Ruditskiy, Y. Xia, 25th anniversary article: galvanic replacement: a simple and versatile route to hollow nanostructures with tunable and well-controlled properties, Adv. Mater. 25 (2013) 6313–6333, https://doi.org/10.1002/adma.201302820.

[82] Q. Chen, Bi-continuous nanoporous structure formation via compound decomposition, J. Electrochem. Soc. 161 (2014) H643–H646, https://doi.org/10.1149/2.0571410jes.

[83] C. Wang, G. Zhu, P. Liu, Q. Chen, Monolithic nanoporous Zn anode for rechargeable alkaline batteries, ACS Nano 14 (2020) 2404–2411, https://doi.org/10.1021/acsnano.9b09669.

[84] C. Wang, Q. Chen, Reduction-induced decomposition: spontaneous formation of monolithic nanoporous metals of tunable structural hierarchy and porosity, Chem. Mater. 30 (2018) 3894–3900, https://doi.org/10.1021/acs.chemmater.8b01431.

[85] W. Xiao, X. Jin, Y. Deng, D. Wang, G.Z. Chen, Three-phase interlines electrochemically driven into insulator compounds: a penetration model and its verification by electroreduction of solid AgCl, Chem. Eur J. 13 (2007) 604–612, https://doi.org/10.1002/chem.200600172.

[86] S. Chatterjee, A. Anikin, D. Ghoshal, J.L. Hart, Y. Li, S. Intikhab, D.A. Chareev, O.S. Volkova, A.N. Vasiliev, M.L. Taheri, N. Koratkar, G. Karapetrov, J. Snyder, Nanoporous metals from thermal decomposition of transition metal dichalcogenides, Acta Mater. 184 (2020) 79–85, https://doi.org/10.1016/j.actamat.2019.11.018.

[87] L. Lu, P. Andela, J. Th, M. De Hosson, Y. Pei, Template-free synthesis of nanoporous nickel and alloys as binder-free current collectors of Li ion batteries, ACS Appl. Nano Mater. 1 (2018) 2206–2218, https://doi.org/10.1021/acsanm.8b00284.

[88] S.-H. Joo, T. Wada, H. Kato, Development of porous FeCo by liquid metal dealloying: evolution of porous morphology and effect of interaction between ligaments and melt, Mater. Des. 180 (2019) 107908, https://doi.org/10.1016/j.matdes.2019.107908.

[89] T. Wada, T. Ichitsubo, K. Yubuta, H. Segawa, H. Yoshida, H. Kato, Bulk-nanoporous-silicon negative electrode with extremely high cyclability for lithium-ion batteries prepared using a top-down process, Nano Lett. 14 (2014) 4505–4510, https://doi.org/10.1021/nl501500g.

[90] T. Wada, K. Yubuta, A. Inoue, H. Kato, Dealloying by metallic melt, Mater. Lett. 65 (2011) 1076–1078, https://doi.org/10.1016/j.matlet.2011.01.054.

CHAPTER FOUR

Monolithic bulk nanoporous zinc by free corrosion dealloying

4.1 Overview

The basic question arises as to what the ideal format of nanoporous metal fuels actually might be, i.e., metal powder versus a monolithic piece of metal. The format of nanoporous metal fuels is critical for at least two reasons: *processability* and *reactivity*. Starting with the second reason, reactivity, so far, we have shown that a nanoporous metal particle reacts more effectively with an oxidizer, such as H_2O and O_2, than a dense metal particle (we refer to Fig. 2.1 in Chapter 2). This is because the effective surface area of the metal exposed to the oxidizer is higher in a nanoporous particle than in a dense metal particle. Using the same rationale, one can show that nanoporous metal in powder form will react more effectively with the oxidant than a monolithic piece of nanoporous metal. The former reason, processability, is related to the fabrication of nanoporous metal fuels by dealloying. We will show that the fabrication of nanoporous metals in powder form is more scalable than making monolithic bulk nanoporous metals. This is because the fabrication of nanoporous metals by dealloying involves chemical reactions (see Chapter 3), which in nature, are kinetically sluggish in bulk materials. As a consequence, making bulk nanoporous metals by dealloying is slower than making nanoporous metals in the powder form. In Chapters 4 and 5, we will use monolithic bulk nanoporous Zn (Chapter 4) and nanoporous Zn powder (Chapter 5) as a model system to elucidate the two reasons mentioned above, i.e., *processability* (making nanoporous Zn) and *reactivity* (reacting nanoporous Zn with H_2O as the oxidant). Similarly, in Chapters 6 and 7, we will use monolithic bulk nanoporous Al (Chapter 6) and nanoporous Al powder (Chapter 7) as another model system to further support our arguments.

Metals as Clean Fuels
ISBN: 978-0-443-13537-8
https://doi.org/10.1016/B978-0-443-13537-8.00003-8

4.2 Fundamental barriers to the synthesis of nanoporous zinc by dealloying

This chapter will focus on the fabrication of monolithic bulk nanoporous Zn and its reactivity with H_2O to produce hydrogen. In general, the relatively high chemical reactivity of nanoporous metals made of nonprecious elements makes them attractive for specific applications where high chemical reactivity is desirable [1−28]. However, the same high chemical reactivity hinders the synthesis of nonprecious nanoporous metals by dealloying. From an electrochemistry viewpoint, the standard reduction potential (SRP) of an element is a good measure of its reactivity with H_2O. The more negative the SRP versus the standard hydrogen electrode (SHE), the more reactive it is with H_2O. The SRPs of most nonprecious elements are very negative (i.e., far below the SHE). As a consequence, common dealloying methods to make nanoporous metals by selective alloy corrosion in aqueous media are more suitable for fabricating precious nanoporous metals, which exhibit a relatively low chemical reactivity compared to nonprecious metals [1−28]. This is because the SRP of most precious metals is much higher than, or close to, the SHE, so that during the fabrication of precious nanoporous metal by dealloying in an aqueous acidic or alkaline corroding medium, the sacrificial alloy component with a more negative SRP is selectively dissolved from the parent alloy, while the corroding medium does not attack the more precious component with positive or less negative SRP (we refer to Chapter 3 for details). However, during the fabrication of a nonprecious nanoporous metal from a parent alloy in which the SRPs of all the alloy components are very negative, the two alloy components are susceptible to reacting with the aqueous corroding medium, causing the parent alloy to either fully dissolve in the corroding medium or undergo passivation [28,29]. Alloy passivation will automatically block the selective corrosion process [30]. These fundamental barriers have hindered the fabrication of nonprecious nanoporous materials from aqueous corroding media.

Among nonprecious metals, Zn with an SRP of −0.76 V versus SHE is susceptible to spontaneously reacting with an aqueous acidic and alkaline corroding medium, which makes the fabrication of nanoporous Zn (NP−Zn) rather complicated [31,32]. However, here we will demonstrate that monolithic bulk NP-Zn with bulk dimensions in the centimeter range can be synthesized in aqueous media using a properly designed parent material, in combination with a good control of the pH of the corroding medium. Next, we will discuss the reactivity of the synthesized NP-Zn as a fuel, with H_2O as the oxidizer (i.e., $Zn-H_2O$ reaction).

4.3 Chemical reaction mechanisms for the synthesis of nanoporous zinc by free corrosion dealloying

In this section, we elucidate the chemical reaction mechanisms associated with the fabrication of NP-Zn by free corrosion dealloying. As the starting point, note that an aqueous corroding solution that cannot dissolve Zn is necessary to make NP-Zn by free corrosion dealloying. In general, metallic Zn can dissolve in aqueous corroding media through the following oxidation reaction, Eq. (4.1):

$$Zn\ (s) \rightarrow Zn^{2+}\ (aq) + 2e^{-} \qquad (4.1)$$

This reaction requires at least −0.76 V versus SHE and will only proceed if it is coupled to a reduction counter-reaction that can generate the required potential [21], as discussed in detail in Chapter 3. This reduction counter-reaction involves either protons (H^{+}) reduction (in the case of a nonoxidizing acidic medium) or water reduction (in the case of an alkaline medium) into H_2 gas through Eqs. (4.2) and (4.3), respectively [33]:

$$2H^{+}\ (aq) + 2e^{-} \rightarrow H_2\ (g) \qquad (4.2)$$

$$2H_2O\ (l) + 2e^{-} \rightarrow H_2\ (g) + 2\ OH^{-}\ (aq) \qquad (4.3)$$

Chapter 3 shows (see Fig. 3.10) that the counter reactions in Eqs. (4.2) and (4.3) are pH-dependent, and the generated pH-dependent voltage is given by Eq. (4.4) [33]:

$$-59.2\ \text{mV} \times \text{pH} \qquad (4.4)$$

In the situation where the pH is null, Eq. (4.2) will generate 0.0 V versus SHE, meaning there is a very high overpotential of +0.76 V (i.e., 0.0 V minus −0.76 V) available to drive the dissolution of Zn through Eq. (4.1) in combination with the reduction counter-reaction in Eq. (4.2). Thus, NP-Zn cannot be fabricated in a corroding medium with pH = 0. In an alkaline solution, for instance, when the pH = 12, there is still an overpotential of ~50 mV (i.e., −0.71 V minus −0.76 V) available to drive the oxidation of Zn through Eq. (4.1), this time in combination with Eq. (4.3). This may result in Zn dissolution or the formation of ZnO (i.e., passivation). In fact, the Pourbaix diagram of Zn predicts that Zn transforms into ZnO in the pH range between 8 and 12, approximately. Thus, even in such an alkaline solution (pH = 12), NP-Zn cannot be fabricated by Zn oxidation through Eq. (4.1). However, a careful analysis of the potential-pH diagram of a nonoxidizing aqueous corroding medium suggests that there is a small pH window (highlighted in the green region in Fig. 3.10) where Zn should not dissolve rapidly through Eq. (4.1) in combination with Eq. (4.3) because there is no driving force to enable that reaction. Indeed, it can be deduced from Eq. (4.4) that when the pH of the corroding medium is around 12.9 (or higher), there is no driving overpotential for Zn dissolution (i.e., −0.76 V minus −0.76 V), meaning that NP-Zn can be synthesized in alkaline solution with pH higher than 12.9. Note that controlling the pH is critical, not only in acids but also in alkaline solutions. Indeed, Zn will be converted to Zn oxide in alkaline corroding media with a much higher concentration (e.g., 2 M NaOH), which restricts the upper concentration limit for the corrosion window [29].

Having identified the condition under which Zn dissolution is kinetically restricted, we next address the sacrificial component. The sacrificial element in the parent alloy used to produce NP-Zn should be more reactive than Zn. Metallic Al, with an SRP of −1.66V versus SHE, is a good candidate sacrificial material since it can readily dissolve through the oxidation reaction in Eq. (4.5):

$$Al\ (s) \rightarrow Al^{3+}\ (aq) + 3e^{-} \tag{4.5}$$

In an alkaline solution, the potential of −1.66 V required to drive Al oxidation can be provided by the reduction counter-reaction in Eq. (4.3). While there is no driving overpotential to dissolve Zn through Eq. (4.1) in combination with the reduction counter-reaction in Eq. (4.3) at the pH of 12.9 or higher, as discussed above, under similar conditions (pH = 12.9), there is a very high overpotential of ~0.90 V available for Al dissolution (i.e., −0.76 V minus −1.66 V). Therefore, if Al is used as the sacrificial component, it will selectively dissolve, while Zn dissolution is kinetically restricted. Based on these considerations, we used $Zn_{20}Al_{80}$ at. % alloys as parent materials to demonstrate the concept. In the next section, we discussed the fabrication of this parent alloy and the phases present.

4.4 Synthesis of metastable $Zn_{20}Al_{80}$ at. % parent alloy

A monolithic Zn—Al parent alloy with a composition of $Zn_{20}Al_{80}$ at. % was fabricated by melting pure Zn and pure Al metals with the appropriate ratio. In a typical synthesis process, ~2.2 g of pure Zn and ~3.5 g of pure Al were melted at 800°C in a graphite boat using a quartz tube under argon flow. The resulting ingot was cooled down, flipped, and remelted twice to ensure a homogeneous mixture. Attempts to dealloy as-prepared bulk $Zn_{20}Al_{80}$ at. % alloy did not result in the formation of NP-Zn. This is because of the coexistence of two phases with significantly different Al contents in $Zn_{20}Al_{80}$ at. % alloys, namely, an Al-rich α phase (FCC) and a Zn-rich η phase (HCP). The maximum Al content in the Zn-rich phase is far below 5 at. % at room temperature (see Fig. 4.1a [34]). This high difference in the sacrificial Al content between the α and η phases results in a high difference in the rate at which Al is etched from these two phases during dealloying at room temperature. Simply put, the α phase can be dealloyed, but the η phase cannot.

A similar behavior (i.e., the coexistence of two phases with different dealloying rates in the parent alloys) has been reported in Ag_xAl_{1-x} parent alloys used to make nanoporous Ag. These two phases were an Al-rich phase and an Ag-rich phase. It was found that dealloying at high temperatures was necessary to remove Al from the Ag-rich phase [35]. Instead of dealloying at high temperatures, Detsi and DeHosson used quenched or metastable Ag_xAl_{1-x} parent alloys to overcome this issue in the Ag_xAl_{1-x} system [2]. Similarly, Hayes et al. used quenched Cu_xMn_{1-x} alloys to overcome the same issue in the Cu_xMn_{1-x} system [3]. Therefore, for the present Zn—Al system, the parent alloy with a composition of $Zn_{20}Al_{80}$ at. % was reheated and annealed at 400°C (red dashed line Fig. 4.1a) for several hours, then quenched in an ice bath to suppress the formation of the Zn-rich HCP η phase [2,36]. This quenching process induces the formation of two FCC phases, an α-FCC phase (with Al content above 80 at. %) and a β-FCC phase containing ~50 at. % of Al (see phase diagram in Fig. 4.1a). It should be emphasized that the β-FCC phase contains significantly more Al than the 5 at. % Al present in the nonquenched HCP η phase. In other words, each of these two FCC phases contains enough sacrificial Al to generate a nanoporous Zn architecture during Al removal by dealloying.

Fig. 4.1b shows the X-ray diffraction (XRD) data with the phases present in the $Zn_{20}Al_{80}$ parent alloy as-made (blue pattern) and after annealing

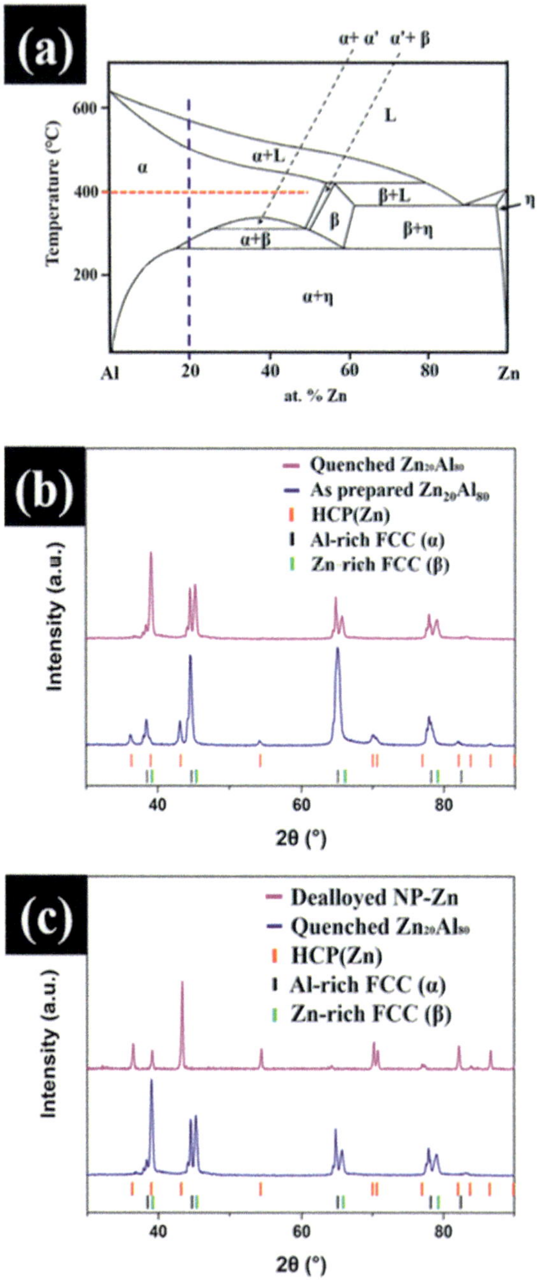

Figure 4.1 (a) Equilibrium phase diagram of Zn—Al binary alloy. The blue dashed line indicates the alloy composition, and the red dashed line shows the annealing temperature for solid-solid transformation to induce the metastable phase after rapid cooling. (b) XRD diffraction patterns of the $Zn_{20}Al_{80}$ at. % parent alloy as-made (blue pattern) and after annealing followed by rapid cooling (purple pattern). (c) XRD diffraction patterns of the metastable $Zn_{20}Al_{80}$ at. % parent alloy before (blue pattern) and after Al removal to make hierarchical nanoporous Zn (purple pattern).

followed by quenching (purple pattern). It is seen that the as-made $Zn_{20}Al_{80}$ parent alloy is a mixture of FCC-Al (*Fm-3m* space group, matching JCPDS card number 00-004-0787) and HCP-Zn (*P63/mmc* space group, matching JCPDS card number 00-004-0831) phases, while the quenched alloy predominantly consists of the FCC phase, with a very small residual of HCP phase as depicted by the peak at $2\theta \sim 84°$ (compare the purple and blue patterns from Fig. 4.1b). The splitting of the FCC peaks (purple pattern in Fig. 4.1b) is associated with the presence of two different FCC phases (α and β as mentioned above). In particular, the diffraction peaks of the quenched Zn-rich β-phase were found to match the Fm-3m(225) space group (PDF card number is #72-1538). Due to the difference in sacrificial Al content in these two FCC phases, we anticipate that removal of Al from the α-FCC phase produces big pores because of the high content of Al in this phase (above 80 at. %), while removal of Al from the β-FCC phase produces small pores because of the relatively lower content of Al in this second phase ($\sim$50 at. %) [37,38].

4.5 Synthesis of monolithic bulk nanoporous Zn by free corrosion dealloying

Monolithic bulk NP-Zn was obtained by free corrosion dealloying, during which Al was selectively removed from the metastable $Zn_{20}Al_{80}$ precursors in alkaline solutions with pH $\sim$14. During dealloying, Al dissolved through the oxidation reaction in Eq. (4.5) in combination with the reduction counter-reaction in Eq. (4.3), as explained earlier in Section 4.3. In a typical synthesis process, the quenched $Zn_{20}Al_{80}$ alloy precursor was first polished with sandpaper to remove possible surface oxide and impurities introduced during arc melting. This parent alloy was then kept in 1 M NaOH solution for 3 days until no H_2 bubbles were observed. Since the pH of the solution is susceptible to change during dealloying, the solution was refreshed every 12 hours to avoid significant changes in the pH. After dealloying, the sample color changed from metallic silver to black/dark gray, and it was easily broken into small fragments, which together indicate the completion of the dealloying process. The sample was then washed with DI water to remove surface ion species (i.e., Na^+ from NaOH solution), dried under an inert environment, and stored in a glove box. It is emphasized that the relatively long dealloying time of 3 days is attributed to the big size of the starting bulk parent alloy (centimeter-sized samples). For large-scale production of NP-Zn, a dealloying process that can be completed in a few minutes, as opposed to a few days, is desirable. In the next chapter, we will demonstrate that NP-Zn in powder form can be produced in less than 5 minutes using micrometer-sized parent alloys.

Fig. 4.2a and b show the fabricated bulk NP-Zn. Large pieces of NP-Zn with bulk dimensions in the range of a few centimeters were made, rinsed in distilled water, dried under an inert environment, and kept in an argon-filled glove box. Due to the high chemical reactivity and air sensitivity of Zn, it is important to confirm that the fabricated material is metallic Zn instead of Zn oxide. This was verified using XRD and energy dispersive spectroscopy (EDS), shown in Figs. 4.1c and 4.3, respectively. The XRD pattern of dealloyed NP-Zn (purple pattern, Fig. 4.1c) matches perfectly with that of metallic Zn (HCP Zn, *P63/mmc* space group, matching JCPDS card number 00-004-0831), which indicates that the synthesized material is not Zn oxide. Note that the dealloyed NP-Zn was crushed into powder and sandwiched between Kapton tape prior to XRD characterization to minimize the exposure of the sample to air. The EDS spectrum in Fig. 4.3 suggests that most of the Al has been removed, and the average residual Al content was below 10 at. %. In addition to EDS, ICP analysis of the dealloyed NP-Zn was conducted to determine the bulk chemical composition. The result suggests that the NP-Zn consists of ~96 at. % of Zn and ~4 at. % of Al. Both ICP and EDS results are summarized in Table 4.1.

It should be noted that the EDS result also indicates the presence of a small oxygen signal. Since material exposure during sample preparation and SEM/EDS characterization is unavoidable, we speculate that this oxygen signal primarily comes from material exposure during sample preparation and characterization. In fact, a similar oxygen signal was observed on the starting $Zn_{20}Al_{80}$ parent alloy (Fig. 4.3). Furthermore, because the dealloying process was carried out in an aqueous alkaline medium containing oxygen species (OH^-, O^{2-}), one expects some residual oxygen species to be present at the surface of NP-Zn after dealloying. Note, however, that since the fabricated NP-Zn was found to be highly reactive with air, possible surface coverage with residual oxygen species cannot be conformal; otherwise, the dealloyed NP-Zn would be passivated and not air-sensitive. Instead, we found that NP-Zn, fabricated in aqueous solutions, is highly air-sensitive: when an NP-Zn sample (initially kept in an argon-filled glove box) was taken from the argon environment to air for materials characterization, it violently reacted with oxygen and moisture from the air, generating enough heat to readily melt a polystyrene weighing boat (melting point 70°C), as seen in Fig. 4.2c, which proves the very "reactive" nature of

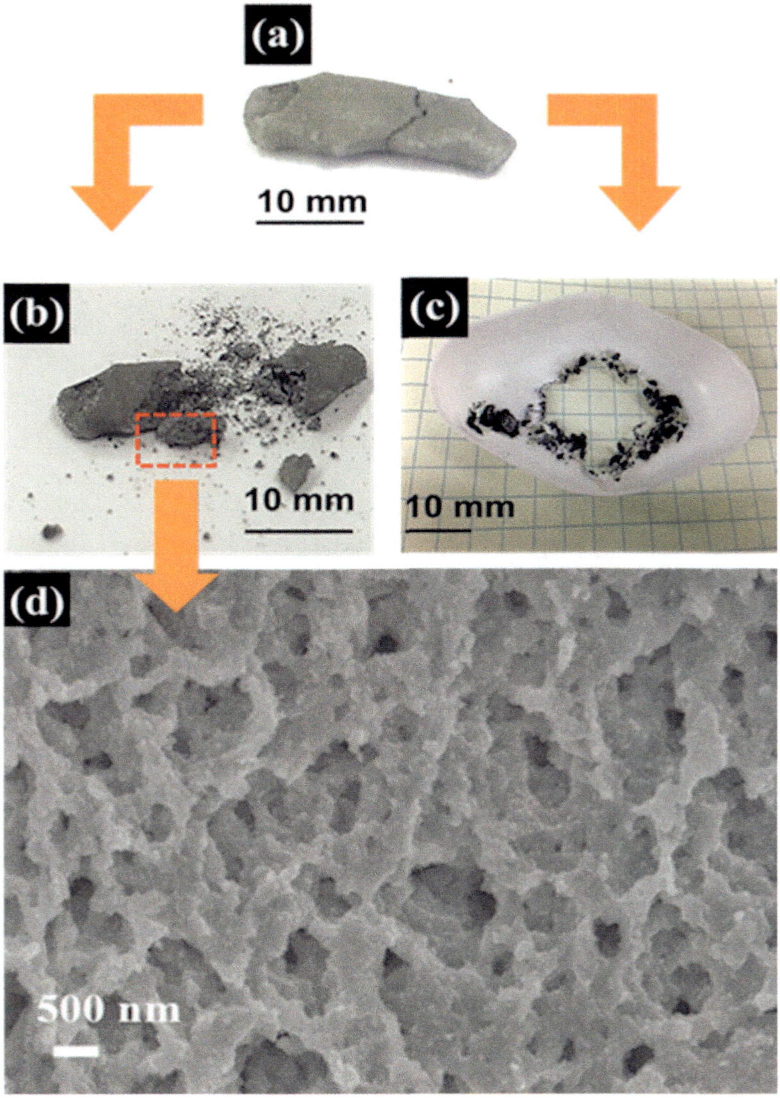

Figure 4.2 (a) Bulk NP-Zn. (b) The bulk piece of NP-Zn is intentionally broken into small fragments to demonstrate its brittleness. (c) Broken pieces of NP-Zn violently react with water moisture from air, generating enough heat to readily melt a polystyrene weighing boat. (d) SEM image of dealloyed NP-Zn.

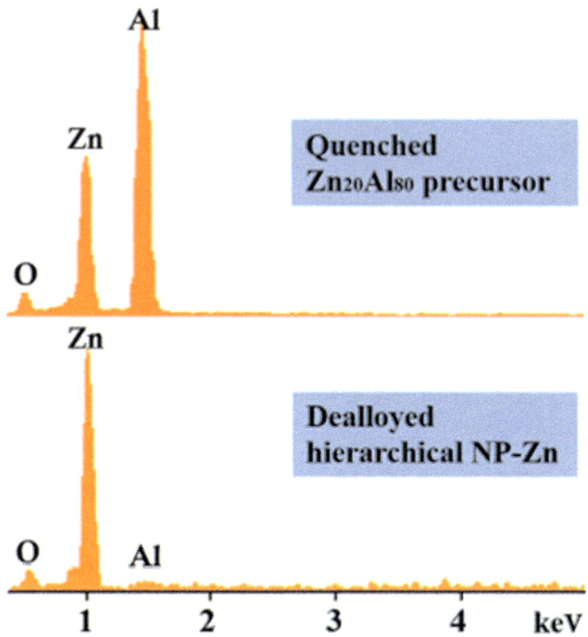

Figure 4.3 EDS spectra of $Zn_{20}Al_{80}$ at. % before (above) and after (below) dealloying. The strong Al signal in the starting parent alloy almost entirely vanishes after dealloying.

Table 4.1 ICP and EDS data on the chemical composition of dealloyed bulk NP-Zn.

	Zn (at. %)	Al (at. %)	Percentage of Al extracted from the Zn–Al parent alloy
EDS analysis	>90	<10	>97.2
ICP analysis[a]	95.8	4.2	98.9

[a]NP-Zn (with residual Al) was dissolved in a mixture of HNO_3 and HCl for ICP analysis.

this class of materials when they are in the nanostructured form and also proves its attractiveness for use as metal fuels.

The scanning electron micrograph (SEM) in Fig. 4.2d shows the typical morphology of the fabricated NP-Zn. It exhibits a hierarchical porous architecture further displayed at different magnifications in Fig. 4.4, where small ligament/pore structures with average size in the range of 10–20 nm are observed, together with large ligament/pore structures with average size in the submicrometer range, i.e., 200–500 nm. TEM images in Fig. 4.5 confirm that the big pores are in the submicrometer range, while the finer

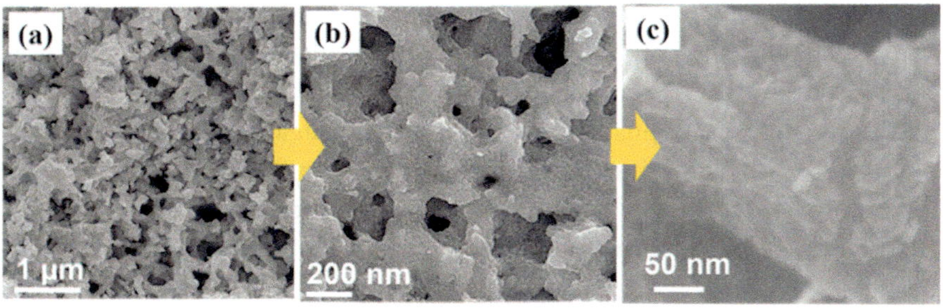

Figure 4.4 SEM images showing the typical hierarchical NP-Zn architecture at different magnifications. (a) Low magnification and (b) intermediate magnification, where large ligament/pore structures with an average size in the range of 200–500 nm can be seen. (c) Under high magnifications, small ligament/pore structures with an average size in the range of 10–20 nm are observed.

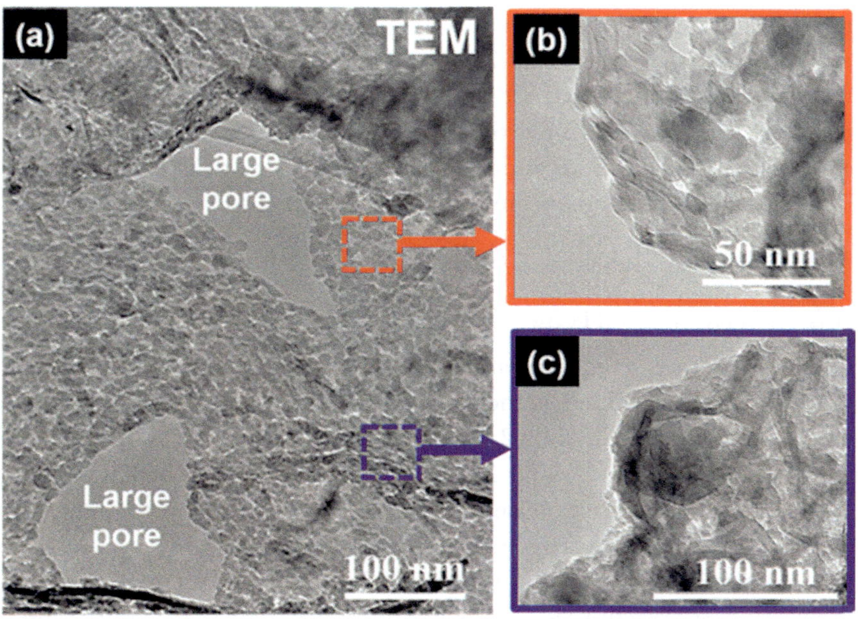

Figure 4.5 TEM images showing the typical hierarchical NP-Zn architecture at different magnifications. (a) At low magnifications, large pores (of the order of 100–300 nm) can be seen. (b, c) Under high magnifications, it is found that the "large" ligaments themselves are made of smaller ligament/pore structures with an average size in the range of 10–20 nm.

pore/ligament sizes are 10–20 nm. The origin of hierarchical porous architecture is attributed to the presence of two FCC phases with different sacrificial Al contents in the starting metastable $Zn_{20}Al_{80}$ parent alloy. Removal of Al from the α FCC phase gives rise to the big ligament/pore structures, due to the relatively high content of Al (more than 80 at. %) in this phase, while removal of Al from the β FCC phase produces small ligament/pore structures due to the relatively lower content of Al ($\sim$50 at. %) in this second phase [37].

4.6 Enhanced reactivity of monolithic nanoporous Zn fuel with water as the oxidizer

The objective of this section is to demonstrate the enhanced reactivity of the dealloyed NP-Zn for use as a metal fuel. In doing so, we will use deionized (DI) neutral H_2O as the oxidizer. The use of DI H_2O is to ensure that there are no ions in the water that will enhance the reactivity of NP-Zn. As discussed in Chapters 1 and 2, activated Zn metal spontaneously reacts with water to produce H_2, heat, and the corresponding metal oxide or hydroxide via one of the two reactions in Eq. (1.12) and Eq. (4.6) below. The nature of the solid reaction product (i.e., $Zn(OH)_2$ or ZnO) depends on the reaction temperature and/or pressure; see also Eq. (1.12):

$$Zn\ (s) + H_2O\ (l) \rightarrow H_2\ (g) + ZnO\ (s) + \text{heat} \tag{1.12}$$

$$Zn\ (s) + 2H_2O\ (l) \rightarrow H_2\ (g) + Zn(OH)_2\ (s) + \text{heat} \tag{4.6}$$

The theoretical gravimetric amount of H_2 produced from the Zn–H_2O reaction, which is defined as the total mass of H_2 over the stoichiometric mass of Zn and H_2O used, is 2.4 wt.% if the Zn–H_2O reaction proceeds through Eq. (1.12), and about 2.0 wt.% if it proceeds through Eq. (4.6). Thus, the Zn–H_2O reaction can be used to gain insight into the reactivity of NP-Zn: the faster hydrogen is produced, the more effective the Zn–H_2O reaction. We know from Chapter 2 that a fundamental challenge with metal-H_2O reactions is the self-liming nature of these reactions, caused by the metal oxide or hydroxide solid by-product. Indeed, recall that during the metal-H_2O reaction, the metal oxide or hydroxide solid by-product forms a layer around the metal, which prevents water from coming into contact with the pure metal underneath to further react, blocking in that way the overall metal-H_2O reaction. Because of this self-limiting issue, most often, metal-H_2O reactions will not proceed when bulk metals are used [39–43]. This justifies the need for activating metal fuels, for example, through nanostructuring, to make nanoporous metals that are more reactive than their dense counterparts (see Chapter 2).

Therefore, we want to verify that the Zn–H_2O reaction can proceed to a certain extent when bulk NP-Zn is used instead of dense Zn. We do so by using a customized gas chromatography (GC) system to investigate the amount and rate of H_2 produced from the reaction between NP-Zn and H_2O. The schematic of our reaction chamber is shown in Fig. 4.6a. Helium gas was used as a carrier to transfer the produced H_2 from the reaction chamber to the GC. The three curves in Fig. 4.6b represent typical raw GC data associated with the amount of H_2 produced at random times of 65, 185, and 260 minutes after introducing H_2O in the reaction chamber. As time increases, the cumulative amount of H_2 generated increases. Fig. 4.6c shows the cumulative amount of hydrogen produced as a function of total reaction time (minutes). The H_2 generation rate was found to be constant over 8 hours, as depicted by the linear fit (red line) in Fig. 4.6c. From the slope of that linear fit, we derived the average hydrogen generation rate of 7.26 μg/minute per gram of Zn, which corresponds to 0.08 mL/minute per gram of Zn. So far, several water-reactive materials, including Al, Zn, and Si, have been used to produce H_2 through respective metal-H_2O reactions [42–45]. To highlight the performance of our bulk NP-Zn toward hydrogen generation, we have compared the reactivity of our bulk NP-Zn with other forms of metal Zn including commercial Zn powder and commercial Zn foil. We have found that only our bulk NP-Zn can react with neutral water to produce hydrogen. Other forms of Zn do not react with neutral water due to the presence of native Zn oxide. Even if the native oxide present on commercial bulk Zn is removed, the reaction of bulk Zn with water will not continually proceed because of the absence of nanoscale features. In fact, as discussed in Chapter 2, several approaches were investigated to overcome this issue of the water-blocking oxide/hydroxide layer in water-reactive metals, including chemical removal of the oxide/hydroxide layer through additives in water [46–49]; chemical removal of the oxide/hydroxide layer through additives/promoters in the metal or use of alloys instead of pure metal [39–43]; and mechanical removal of the oxide/hydroxide layer through frictions and milling [50,51]. However, these solutions will add "non-active" mass to the system—through the use of a catalyst or a nonreactive alloying metal—and thus lower its theoretical gravimetric hydrogen storage capacity. Nevertheless, our hierarchical NP-Zn can generate hydrogen continuously over 8 hours due to the ultrafine size of

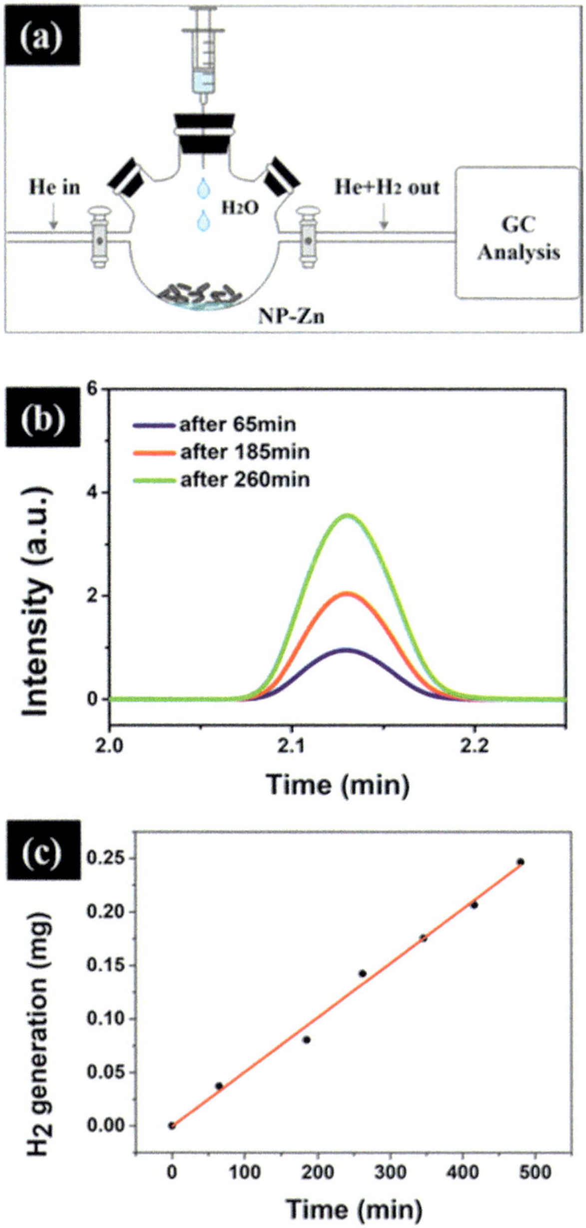

Figure 4.6 (a) Illustration of the setup used to measure the amount and rate of H_2 generated from NP-Zn in contact with distilled H_2O. Generated H_2 is pushed into the GC using helium gas as a carrier. (b) Typical GC raw data showing H_2 peak intensity versus reaction time. (c) Cumulative amount of H_2 produced versus reaction time. The straight line is the corresponding linear fitting.

NP-Zn. Indeed, it is less likely that such small structure sizes (10–20 nm) with irregular shapes and random curvatures can be conformally covered with Zn oxide/hydroxide to block the reaction of Zn with water by preventing water from coming into contact with fresh nanostructured Zn. In addition, while the small structure size (10–20 nm) provides a high surface area for the reaction with water, the big ligaments and pores (200–500 nm) in the hierarchical NP-Zn structure facilitate the reaction kinetics by effectively facilitating mass transport of the reactant (H_2O) and product (H_2) in and out of the bulk material. H_2 measurements were stopped after ~8 hours while the system was still producing H_2 at a constant rate (see Fig. 4.6c).

On the other hand, the amount of H_2 collected after the reaction has fully completed reaches ~16% of the theoretical gravimetric capacity of the Zn/H_2O system. The relatively low capacity is caused by two factors: (i) the bulk nature of the monolithic NP-Zn. Indeed, we will show in Chapter 4 that using NP-Zn in powder form significantly increases the reaction yield. (ii) Metallic Zn is, in nature, a very poor H_2 evolution catalyst. As such, the surface of Zn does not promote H_2 evolution. Decorating the surface of NP-Zn with atom clusters of an H_2 evolution catalyst, such as Ni clusters as illustrated in Fig. 2.1b–d (from Chapter 2), will significantly enhance the yield of the reaction.

4.7 Conclusions

Nanoporous metals made of nonprecious elements are highly chemically reactive, which makes them appealing for applications where high reactivity is needed, such as using them as fuels. However, this same high chemical reactivity makes it difficult to synthesize nonprecious nanoporous metals through dealloying. In this chapter, we discussed the creation of monolithic bulk nanoporous Zn and its reactivity with H_2O to generate hydrogen.

First, we elucidate the fundamental reaction mechanisms associated with the fabrication of nanoporous Zn by free corrosion dealloying in an aqueous solution and identify the pH range at which Zn dissolution is kinetically restricted, compared to the dissolution rate of a sacrificial element like Al. Based on these considerations, we used $Zn_{20}Al_{80}$ at. % alloys as parent materials to fabricate nanoporous Zn through a novel pH-controlled dealloying method from quenched $Zn_{20}Al_{80}$ at. % parent alloys. These quenched $Zn_{20}Al_{80}$ at. % parent alloys consist of two face-centered cubic phases (α and β) with varying Al content, leading to a hierarchical nanoporous architecture upon the removal of Al from each phase. It was found that the α-fcc phase forms comparatively larger ligament/pore structures (200—500 nm) as a result of its high Al content (over 80 at. %), whereas the β-fcc phase leads to smaller ligament/pore structures (10—20 nm) due to its lower aluminum content (approximately 50 at. %).

Finally, we demonstrate the increased reactivity of dealloyed nanoporous Zn for its potential use as a wet metal fuel. In our investigation, we study the reactivity of nanoporous Zn with deionized H_2O to produce hydrogen. We utilize deionized H_2O as the oxidizer to eliminate any ions in the water that could potentially boost the reactivity of the nanoporous Zn. Due to the small feature sizes in this hierarchical structure, the nanoporous Zn can continuously split water into H_2 and Zn $(OH)_2$ for 8 hours at a steady rate of 0.08 mL/minute per gram of Zn. We argue that the macropores streamline the movement of water into the bulk interior of the alloy, whereas the mesopores offer an extensive internal surface area that allows the H_2O splitting reaction to occur efficiently. Needless to say, producing hydrogen from water-reactive metals is appealing for numerous applications, such as on-site H_2 generation for powering hydrogen fuel cell equipment.

Acknowledgments

Part of the content of this chapter has been published by J. Fu, Z. Deng, T. Lee, J.S. Corsi, Z. Wang, D. Zhang, and E. Detsi, ACS Applied Energy Materials 1 (7), pp 3198—3205 (2018), entitled *pH-Controlled Dealloying Route to Hierarchical Bulk Nanoporous Zn Derived from Metastable Alloy for Hydrogen Generation by Hydrolysis of Zn in Neutral Water*. The authors are thankful to Jintao Fu, Ziling Deng, Timothy Lee, John Corsi, Zeyu Wang, and Dongyang Zhang for their discussion and contributions.

References

[1] J. Erlebacher, M.J. Aziz, A. Karma, N. Dimitrov, K. Sieradzki, Evolution of nanoporosity in dealloying, Nature 410 (2001) 450—453, https://doi.org/10.1038/35068529.

[2] E. Detsi, Z. Vuković, S. Punzhin, P.M. Bronsveld, P.R. Onck, J.T.M.D. Hosson, Fine-tuning the feature size of nanoporous silver, CrystEngComm 14 (2012) 5402—5406, https://doi.org/10.1039/c2ce25313e.

[3] J.R. Hayes, A.M. Hodge, J. Biener, A. V Hamza, K. Sieradzki, Monolithic nanoporous copper by dealloying Mn—Cu, J. Mater. Res. 21 (2006) 2611—2616, https://doi.org/10.1557/jmr.2006.0322.

[4] J.B. Cook, E. Detsi, Y. Liu, Y.-L. Liang, H.-S. Kim, X. Petrissans, B. Dunn, S.H. Tolbert, Nanoporous tin with a granular hierarchical ligament morphology as a highly stable Li-ion battery anode, ACS Appl. Mater. Interfaces 9 (2017) 293—303, https://doi.org/10.1021/acsami.6b09014.

[5] F. Meng, Y. Ding, Sub-micrometer-thick all-solid-state supercapacitors with high power and energy densities, Adv. Mater. 23 (2011) 4098—4102, https://doi.org/10.1002/adma.201101678.

[6] X. Lang, A. Hirata, T. Fujita, M. Chen, Nanoporous metal/oxide hybrid electrodes for electrochemical supercapacitors, Nat. Nanotechnol. 6 (2011) 232–236, https://doi.org/10.1038/nnano.2011.13.

[7] M. Hakamada, M. Mabuchi, Fabrication of nanoporous palladium by dealloying and its thermal coarsening, J. Alloys Compd. 479 (2009) 326–329, https://doi.org/10.1016/j.jallcom.2008.12.078.

[8] T.S. Dorofeeva, E. Seker, Electrically tunable pore morphology in nanoporous gold thin films, Nano Res. 8 (2015) 2188–2198, https://doi.org/10.1007/s12274-015-0726-x.

[9] E. Detsi, S. Punzhin, J. Rao, P.R. Onck, J.T.M. De Hosson, Enhanced strain in functional nanoporous gold with a dual microscopic length scale structure, ACS Nano 6 (2012) 3734–3744, https://doi.org/10.1021/nn300179n.

[10] A. Wittstock, V. Zielasek, J. Biener, C.M. Friend, M. Bäumer, Nanoporous gold catalysts for selective gas-phase oxidative coupling of methanol at low temperature, Science 327 (2010) 319–322, https://doi.org/10.1126/science.1183591.

[11] J. Weissmüller, K. Sieradzki, Dealloyed nanoporous materials with interface-controlled behavior, MRS Bull. 43 (2018) 14–19, https://doi.org/10.1557/mrs.2017.299.

[12] E.T. Lilleodden, P.W. Voorhees, On the topological, morphological, and microstructural characterization of nanoporous metals, MRS Bull. 43 (2018) 20–26, https://doi.org/10.1557/mrs.2017.303.

[13] A. Mortari, A. Maaroof, D. Martin, M.B. Cortie, Mesoporous gold electrodes for sensors based on electrochemical double layer capacitance, Sens. Actuat. B Chem. 123 (2007) 262–268, https://doi.org/10.1016/j.snb.2006.08.018.

[14] X. Wang, W. Wang, Z. Qi, C. Zhao, H. Ji, Z. Zhang, Electrochemical catalytic activities of nanoporous palladium rods for methanol electro-oxidation, J. Power Sources 195 (2010) 6740–6747, https://doi.org/10.1016/j.jpowsour.2010.03.098.

[15] E. Detsi, M. Salverda, P.R. Onck, J.T.M. De Hosson, On the localized surface plasmon resonance modes in nanoporous gold films, J. Appl. Phys. 115 (2014) 044308, https://doi.org/10.1063/1.4862440.

[16] E. Detsi, P. Onck, J.T.M. De Hosson, Metallic muscles at work: high rate actuation in nanoporous gold/polyaniline composites, ACS Nano 7 (2013) 4299–4306, https://doi.org/10.1021/nn400803x.

[17] X. Wang, W. Wang, Z. Qi, C. Zhao, H. Ji, Z. Zhang, High catalytic activity of ultrafine nanoporous palladium for electro-oxidation of methanol, ethanol, and formic acid, Electrochem. Commun. 11 (2009) 1896–1899, https://doi.org/10.1016/j.elecom.2009.08.011.

[18] H.J. Jinz, D. Kramer, Y. Ivanisenko, J. Weissmüller, Macroscopically strong nanoporous Pt prepared by dealloying, Adv. Eng. Mater. 9 (2007) 849–854, https://doi.org/10.1002/adem.200700177.

[19] Q. Lu, J. Rosen, Y. Zhou, G.S. Hutchings, Y.C. Kimmel, J.G. Chen, F. Jiao, A selective and efficient electrocatalyst for carbon dioxide reduction, Nat. Commun. 5 (2014) 3242, https://doi.org/10.1038/ncomms4242.

[20] Y.-C. Hsieh, S.D. Senanayake, Y. Zhang, W. Xu, D.E. Polyansky, Effect of chloride anions on the synthesis and enhanced catalytic activity of silver nanocoral electrodes for CO_2 electroreduction, ACS Catal. 5 (2015) 5349–5356, https://doi.org/10.1021/acscatal.5b01235.

[21] Z. Deng, E. Detsi, Enhancing the free corrosion dealloying rate with a catalytically driven reaction, Nanoscale 9 (2017) 11858–11863, https://doi.org/10.1039/c7nr04611a.

[22] M. Hakamada, M. Mabuchi, Preparation of nanoporous Ni and Ni-Cu by dealloying of rolled Ni-Mn and Ni-Cu-Mn alloys, J. Alloys Compd. 485 (2009) 583–587, https://doi.org/10.1016/j.jallcom.2009.06.031.

[23] J.B. Cook, T.C. Lin, E. Detsi, J.N. Weker, S.H. Tolbert, Using X-ray microscopy to understand how nanoporous materials can be used to reduce the large volume change in alloy anodes, Nano Lett. 17 (2017) 870, https://doi.org/10.1021/acs.nanolett.6b04181.

[24] J. Rosen, G.S. Hutchings, Q. Lu, R.V. Forest, A. Moore, F. Jiao, Electrodeposited Zn dendrites with enhanced CO selectivity for electrocatalytic CO_2 reduction, ACS Catal. 5 (2015) 4586–4591, https://doi.org/10.1021/acscatal.5b00922.

[25] F. Quan, D. Zhong, H. Song, F. Jia, L. Zhang, A highly efficient zinc catalyst for selective electroreduction of carbon dioxide in aqueous NaCl solution, J. Mater. Chem. A 3 (2015) 16409–16413, https://doi.org/10.1039/C5TA04102C.

[26] D.H. Won, H. Shin, J. Koh, J. Chung, H.S. Lee, H. Kim, S.I. Woo, Highly efficient, selective, and stable CO_2 electroreduction on a hexagonal Zn catalyst, Angew. Chem. Int. Ed. 55 (2016) 9297–9300, https://doi.org/10.1002/anie.201602888.

[27] Y. Liang, R. Feng, S. Yang, H. Ma, J. Liang, J. Chen, Rechargeable Mg batteries with graphene-like MoS_2 cathode and ultrasmall Mg nanoparticle anode, Adv. Mater. 23 (2011) 640–643, https://doi.org/10.1002/adma.201003560.

[28] Z. Zhang, Y. Wang, Z. Qi, W. Zhang, J. Qin, J. Frenzel, Generalized fabrication of nanoporous metals (Au, Pd, Pt, Ag, and Cu) through chemical dealloying, J. Phys. Chem. C 113 (2009) 12629–12636, https://doi.org/10.1021/jp811445a.

[29] C. Xu, R. Wang, Y. Zhang, Y. Ding, A general corrosion route to nanostructured metal oxides, Nanoscale 2 (2010) 906–909, https://doi.org/10.1039/b9nr00351g.

[30] J. Erlebacher, K. Sieradzki, Pattern formation during dealloying, Scripta Mater. 49 (2003) 991–996, https://doi.org/10.1016/S1359-6462(03)00471-8.

[31] M. Abdallah, Ethoxylated fatty alcohols as corrosion inhibitors for dissolution of zinc in hydrochloric acid, Corrosion Sci. 45 (2003) 2705–2716, https://doi.org/10.1016/S0010-938X(03)00107-0.

[32] J. Dobryszycki, S. Biallozor, On some organic inhibitors of zinc corrosion in alkaline media, Corrosion Sci. 43 (2001) 1309–1319, https://doi.org/10.1016/S0010-938X(00)00155-4.

[33] M.G. Walter, E.L. Warren, J.R. McKone, S.W. Boettcher, Q. Mi, E. a Santori, N.S. Lewis, Solar water splitting cells, Chem. Rev. 110 (2010) 6446–6473, https://doi.org/10.1021/Cr1002326.

[34] M. Demirtas, G. Purcek, H. Yanar, Z.J. Zhang, Z.F. Zhang, Effect of chemical composition and grain size on RT superplasticity of Zn-Al alloys processed by ECAP, Lett. Mater. 5 (2015) 328–334, https://doi.org/10.22226/2410-3535-2015-3-328-334.

[35] X. Wang, Z. Qi, C. Zhao, W. Wang, Z. Zhang, Influence of alloy composition and dealloying solution on the formation and microstructure of monolithic nanoporous silver through chemical dealloying of Al-Ag alloys, J. Phys. Chem. C 113 (2009) 13139–13150, https://doi.org/10.1021/jp902490u.

[36] M. Eleani, C. Edil, F. Dalla, C. Rick, M. Scherer, A. Carlos, B.A. Dedavid, Study of the influence of copper and magnesium additions on the microstructure formation of Zn − Al hypoeutectic alloys, J Alloys Comp. 488 (2009) 89–99, https://doi.org/10.1016/j.jallcom.2009.08.125.

[37] Z. Zhang, Y. Wang, Z. Qi, C. Somsen, X. Wang, C. Zhao, Fabrication and characterization of nanoporous gold composites through chemical dealloying of two phase Al–Au alloys, J. Mater. Chem. 19 (2009) 6042–6050, https://doi.org/10.1039/b904052h.

[38] T. Song, M. Yan, Z. Shi, A. Atrens, M. Qian, Creation of bimodal porous copper materials by an annealing-electrochemical dealloying approach, Electrochim. Acta 164 (2015) 288–296, https://doi.org/10.1016/j.electacta.2015.02.217.

[39] V. Rosenband, A. Gany, Application of activated aluminum powder for generation of hydrogen from water, Int. J. Hydrogen Energy 35 (2010) 10898–10904, https://doi.org/10.1016/j.ijhydene.2010.07.019.
[40] X.N. Huang, C.J. Lv, Y.X. Huang, S. Liu, C. Wang, D. Chen, Effects of amalgam on hydrogen generation by hydrolysis of aluminum with water, Int. J. Hydrogen Energy 36 (2011) 15119–15124, https://doi.org/10.1016/j.ijhydene.2011.08.073.
[41] Z. Zhao, X. Chen, M. Hao, Hydrogen generation by splitting water with Al-Ca alloy, Energy 36 (2011) 2782–2787, https://doi.org/10.1016/j.energy.2011.02.018.
[42] H.Z. Wang, D.Y.C. Leung, M.K.H. Leung, M. Ni, A review on hydrogen production using aluminum and aluminum alloys, Renew. Sustain. Energy Rev. 13 (2009) 845–853, https://doi.org/10.1016/j.rser.2008.02.009.
[43] A.V. Ilyukhina, A.S. Ilyukhin, E.I. Shkolnikov, Hydrogen generation from water by means of activated aluminum, Int. J. Hydrogen Energy (2012) 16382–16387, https://doi.org/10.1016/j.ijhydene.2012.02.175.
[44] F. Erogbogbo, T. Lin, P.M. Tucciarone, K.M. Lajoie, L. Lai, G.D. Patki, P.N. Prasad, M.T. Swihart, On-demand hydrogen generation using nanosilicon: splitting water without light, heat, or electricity, Nano Lett. 13 (2013) 451–456, https://doi.org/10.1021/nl304680w.
[45] H.H. Funke, H. Diaz, X. Liang, C.S. Carney, A.W. Weimer, P. Li, Hydrogen generation by hydrolysis of zinc powder aerosol, Int. J. Hydrogen Energy 33 (2008) 1127–1134, https://doi.org/10.1016/j.ijhydene.2007.12.061.
[46] Y.K. Chen, H. Te Teng, T.Y. Lee, H.W. Wang, Rapid hydrogen generation from aluminum-water system by adjusting water ratio to various aluminum/aluminum hydroxide, Int. J. Energy Environ. Eng. 5 (2014) 1–6, https://doi.org/10.1007/s40095-014-0087-3.
[47] G.L. Ma, H. Bin Dai, D.W. Zhuang, H.J. Xia, P. Wang, Controlled hydrogen generation by reaction of aluminum/sodium hydroxide/sodium stannate solid mixture with water, Int. J. Hydrogen Energy 37 (2012) 5811–5816, https://doi.org/10.1016/j.ijhydene.2011.12.157.
[48] L. Soler, A.M. Candela, J. Macanás, M. Muñoz, J. Casado, In situ generation of hydrogen from water by aluminum corrosion in solutions of sodium aluminate, J. Power Sources 192 (2009) 21–26, https://doi.org/10.1016/j.jpowsour.2008.11.009.
[49] L. Soler, A.M. Candela, J. Macanás, M. Muñoz, J. Casado, Hydrogen generation from water and aluminum promoted by sodium stannate, Int. J. Hydrogen Energy 35 (2010) 1038–1048, https://doi.org/10.1016/j.ijhydene.2009.11.065.
[50] S.S. Razavi-Tousi, J.A. Szpunar, Effect of structural evolution of aluminum powder during ball milling on hydrogen generation in aluminum-water reaction, Int. J. Hydrogen Energy 38 (2013) 795–806, https://doi.org/10.1016/j.ijhydene.2012.10.106.
[51] Y. Kanda, S. Kondo, S. Ooya, T. Kobayashi, Y. Uemichi, K. Higashino, M. Sugioka, Green hydrogen production by mechanical mixing of aluminum with water, J. Chem. Eng. Jpn. 44 (2011) 803–808, https://doi.org/10.1252/jcej.10we272.

CHAPTER FIVE

Rapid synthesis of nanoporous zinc in powder form by free corrosion dealloying

5.1 Overview

The scalable synthesis of non-precious nanoporous metals such as nanoporous zinc (NP-Zn), nanoporous iron (NP-Fe), and nanoporous aluminum (NP-Al) is crucial for metal fuel applications and for large-scale production of hydrogen through the reaction between non-precious metals and water. In Chapter 4, we show that the fabrication of bulk NP-Zn by selective removal of Al from sub-centimeter-sized Zn-Al parent alloys (made by tube furnace) through free corrosion dealloying usually takes a few days. In Chapter 5, we demonstrate that this free corrosion dealloying process can be reduced from a few days to 4 minutes simply by using micrometer-sized Zn-Al powder particles with nominal composition $Zn_{40}Al_{60}$ at. % produced by gas-atomization as the parent alloy. Reducing the size of the parent alloy significantly enhances the dealloying rate. Furthermore, Al and Zn are phase-separated in Zn-Al powder particles due to the gas-atomization process, making removing the sacrificial Al phase easy. The fabricated NP-Zn powder exhibits a hierarchical ligament/pore morphology, with tiny structures in the range of 50−60 nm coming from Zn nanoparticle aggregation during dealloying and large structures in the range of 100−200 nm from the removal of the sacrificial Al phase (Fig. 5.1). We study the reactivity of this NP-Zn for wet metal fuel applications by demonstrating that it can spontaneously react with water at near-neutral pH to produce hydrogen and zinc oxide solid by-products with a hydrogen generation yield of ≈52% within 40 minutes.

5.2 Dealloyed nanoporous systems

Due to their high surface-area-to-volume ratio and ample catalytically active sites, such as specific facets, steps, and kinks, dealloyed nanoporous metals are desirable for various applications, including electrocatalysis

Metals as Clean Fuels
ISBN: 978-0-443-13537-8
https://doi.org/10.1016/B978-0-443-13537-8.00015-4

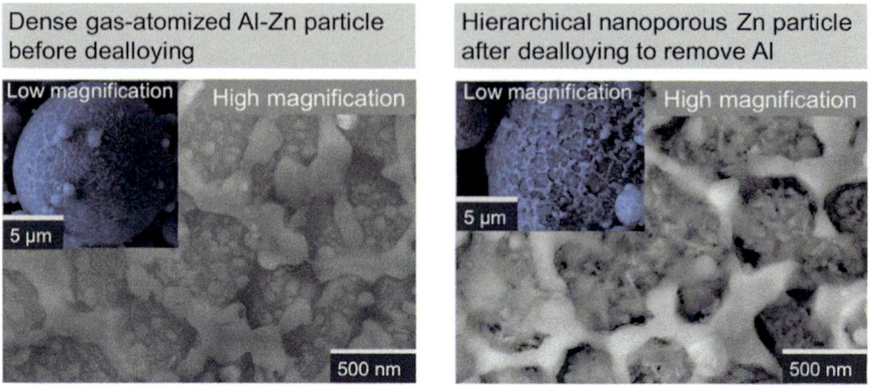

Figure 5.1 The fabricated NP-Zn powder exhibits a hierarchical ligament/pore morphology.

[1,2], energy conversion and storage [3,4], actuation [5,6], plasmonics [7,8], and structural application [9,10]. In particular, bulk nanoporous metals offer the benefits of bulk metals, such as handleability, mechanical integrity, and thermal and electrical conductivity, as well as the benefit of nanostructured materials, such as the size-dependent properties [11], low-density [12], and high catalytic activity [13,14]. In most chemical and electrochemical applications of nanoporous metals, significant focus has been directed toward precious metals, such as nanoporous Pt [15], Pd [16], Ag [17], and Au [18]. This is primarily because the relatively low chemical reactivity of the precious metals makes it easy to fabricate them in the nanoporous format and characterize them. Further, most precious metals are great catalysts as they lower the activation energy barrier and increase the kinetics of the desired reactions [19]. However, the relatively high cost and scarcity of the precious metals hinder their use in large-scale applications, including hydrogen production [20]. Non-precious metals, which are earth-abundant and more affordable, are more attractive for industrial-scale applications than platinum group metals (PGM). Although several studies suggested promising applications for non-precious metals, such as Fe, Zn, Al, and Mg [21—23], non-precious nanoporous metals have seen limited fabrication due to their inherent high chemical reactivity with external environments [1]. Specifically, synthesizing pure, non-precious nanoporous metals is challenging due to their oxygen- and water-sensitive nature. In Chapter 4 and in previous work [24], we demonstrate the successful fabrication of monolithic bulk nanoporous Zn (NP-Zn) by selectively

etching Al from monolithic bulk Zn-Al parent alloys. Compared to dense commercial Zn, this NP-Zn exhibits an enhanced hydrogen generation yield in neutral water. However, the selective leaching process to make the monolithic NP-Zn took over 3 days. The hydrogen generation rate in neutral pH was relatively slow, with the typical hydrogen generation yield less than 20% after 8 hours [24]. For large-scale applications, a process that will enable the synthesis of high-performance NP-Zn in a few minutes, as opposed to a few days, is desirable. Therefore, the present chapter aims to introduce an innovative and scalable method for producing NP-Zn in powder form within 4 minutes instead of 3 days. We verify the high reactivity of the synthesized NP-Zn for use as a wet metal fuel by demonstrating that it can react with water to produce hydrogen with a yield of ≈50% within 30—40 minutes.

5.3 Experimental methods

5.3.1 Dealloying

Commercial Zn-Al parent alloy in powder form, made by gas-atomization, was obtained from Valimet Inc. The nominal composition and the particle size, as provided by the supplier, were $Zn_{40}Al_{60}$ and 1—10 μm, respectively. The Zn-Al alloy powder was used as-received to fabricate NP-Zn by keeping the alloy powder in 1 M NaOH solution for 3, 4, and 5 minutes to remove Al selectively. The solution was stirred at room temperature on a magnetic stirrer plate for uniform leaching. After the Al dissolution, the remaining powder was washed with deionized (DI) water and rinsed with isopropyl alcohol (IPA) for rapid drying. The NP-Zn powder was collected from IPA by vacuum filtration using a Büchner funnel (Fisher) and a Nylon membrane filter with a pore size of 200 nm (Omicron).

5.3.2 Characterization

We used an Xe^{+} FIB/SEM to cut through the powder particles before and after the dealloying to investigate their microstructure. The elemental compositions of the actual Zn-Al parent alloy and the corresponding dealloyed NP-Zn were analyzed with EDS and ICP (Spectro Genesis ICP-OES device). The sample solution for ICP was prepared by dissolving the powders in 0.1 M nitric acid (HNO_3) solution so that the concentrations of the elements are 2—10 ppm, with a standard reference solution (Multielement standard solution five for ICP, Supelco). The crystal structure of the samples before and after the dealloying was investigated using X-ray diffraction (XRD, Rigaku) with Bragg-Brentano para-focusing geometry operating with a Cu Kα_1 line at the wavelength of 1.5405 Å and a horizontal goniometer collecting at an angle range from 10 degrees to 90 degrees with a step size of 0.02 degrees and a scanning speed at 5 degrees/min. Before XRD measurement, the powder sample was sealed inside the glovebox with Kapton tape (Ted Pella Inc.) to minimize air exposure and oxidation. The Zn-Al parent alloy powder and dealloyed NP-Zn samples were stored in an argon environment, with the concentrations of water and oxygen maintained below 0.1 ppm.

5.3.3 Hydrogen generation

The water displacement method (see Section 5.5) was used to measure the hydrogen produced from the reaction between NP-Zn and water at a near-neutral pH. A three-neck flask was used as the reaction chamber for the hydrolysis reaction, and one of the necks was connected to a graduated cylinder flipped and immersed in water. The evolved hydrogen was collected in the cylinder, and the amount of hydrogen was deduced from the volume change in the cylinder. The second neck was used to introduce near-neutral electrolytes, and NP-Zn powder was transferred via the third neck of the flask. Both second and third necks were sealed with glass stoppers during hydrogen production. Vacuum grease and plastic joint clips were applied at all connections to ensure sealings. The flask and the electrolyte were preheated to 65°C in an oil bath to simulate the exothermic behavior during Zn hydrolysis. 50 mg of NP-Zn powder reacted with the 20 mL electrolyte containing 0.01 M NaOH water solution to maintain the pH at 12. The volume change due to hydrogen generation was monitored for 1 hour with 10-minute intervals to record the volume of hydrogen produced, and the total volume change was normalized by the mass of NP-Zn and its stoichiometric amount of hydrogen produced. Each water displacement test was repeated three times to check the consistency, and the average values were plotted for further analysis.

5.4 Nanoporous zinc powder by free corrosion dealloying

The NP-Zn powder was synthesized by free corrosion dealloying. During this process, sacrificial Al is selectively etched away from the parent alloy, which consists of micrometer-sized powder particles with the composition of $Zn_{40}Al_{60}$ at. %, in 1 M NaOH solution (pH of 14). This alkaline environment facilitates the dissolution of the Al, which undergoes an oxidation reaction, as described in Eq. (5.1). Concurrently, a hydrogen evolution counter-reaction, a reduction reaction, occurs as detailed in Eq. (5.2). The electrons released during the oxidation of aluminum (Eq. 5.1) are consumed by the reduction reaction (Eq. 5.2), leading to the formation of the NP-Zn powder. This dealloying process removes the Al, leaving a porous Zn structure behind.

$$Al(s) + 4OH^-(aq) \rightarrow Al(OH)_4^-(aq) + 3e^- \; (E^o = -2.33 \text{ V vs. SHE}) \tag{5.1}$$

$$2H_2O\ (l) + 2e^- \rightarrow H_2(g) + 2OH^-(aq) \; (E^o = -0.83 \text{ V vs. SHE}) \tag{5.2}$$

Interestingly, the Pourbaix diagram in Fig. 5.2a shows that during free corrosion dealloying, besides Al dissolution, Zn should also dissolve in an alkaline solution through the reaction oxidation reaction in Eq. (5.3), combined with the reduction counter-reaction in Eq. (5.2).

$$Zn(s) + 4OH^-(aq) \rightarrow ZnO_2^{2-}(aq) + 2H_2O\ (l) + 2e^- \; (E^o = -1.22 \text{ V vs. SHE}) \tag{5.3}$$

In other words, it may seem impossible to selectively remove Al from the Zn-Al parent alloy without dissolving Zn. Fortunately, due to the high difference in the standard reduction potential of Al ($E^o = -2.33$ V vs. SHE in an alkaline solution) and Zn ($E^o = -1.22$ V vs. SHE in an alkaline solution), a phenomenon known in the corrosion science community as *cathodic protection* will prevent Zn from dissolving. Indeed, when two dissimilar metals with different standard reduction potentials are in physical contact in a corroding environment (Fig. 5.2b), the metal with the lowest standard reduction potential will dissolve with the counter-reaction taking place on the metal with higher standard reduction potential. In the case of Zn-Al alloy, Al will dissolve through Eq. (5.1), and the hydrogen evolution counter-reaction in Eq. (5.2) will preferentially take place on Zn. Since the counter-reaction in Eq. (5.2) is a reduction reaction, it automatically prevents Zn from oxidizing. The dissolution of Zn will commence only after the complete dissolution of all Al. In other words, the samples should not be kept in the corroding medium for a long time; otherwise, Zn will dissolve after Al removal. Therefore, we kept the samples in the corrosion medium

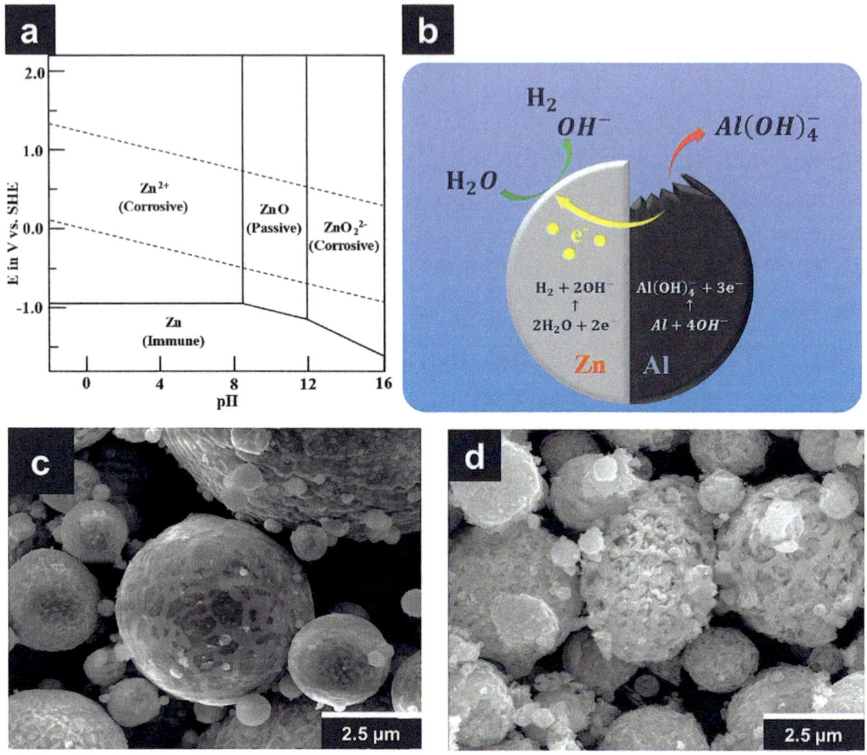

Figure 5.2 (a) Pourbaix diagram of Zn, showing that Zn can dissolve in alkaline when the pH $>$ 12. (b) A simplified free corrosion mechanism of Zn-Al was performed in an alkaline corrosive medium. Zn dissolution is prevented by cathodic protection, as Al oxidation provides electrons to Zn and the coupled reduction counter-reaction. (c) SEM image of Zn-Al parent alloy particles, consisting of two separated phases: Zn and Al. (d) SEM image of NP-Zn, showing the porous structure after dealloying. *(a) Adapted from Ref. [25].*

for 3, 4, and 5 minutes to identify the optimal dealloying time using XRD analysis. The SEM images in Fig. 5.2c and d show the typical microstructure of the powder particles before and after dealloying, respectively. The presence of two phases, a Zn-rich phase, and an Al-rich phase, in the starting Zn-Al parent alloy powder is confirmed by the contrast from the SEM image of Fig. 5.2c. This contrast originates from the difference in the mass/density of the two phases, as the heavier/denser Zn-rich phase deflects electrons more strongly, thus appearing to be lighter. In contrast, the less dense Al-rich phase absorbs more electrons, thus appearing darker (Fig. 5.2c). The NP-Zn powder particles in Fig. 5.2d show that the particle shapes are conserved after leaching the sacrificial phase from the parent alloy, and the dealloying process successfully creates porous powder particles from dense ones.

Fig. 5.3 shows four different XRD patterns for the pristine Zn-Al parent alloy (black) and the corresponding samples after 3 (red), 4 (green), and 5 (orange) minutes of free corrosion dealloying. The pristine alloy powder is a mixture of Al and Zn phases, indicated by the coexistence of both Al and Zn XRD patterns in the black curve in Fig. 5.3. Upon dealloying of the parent alloy powder for 3 minutes, the red XRD patterns in Fig. 5.3 indicate that the resulting product powder is predominantly composed of pure Zn. In addition, minor peaks at 38.5 degrees and 44.8 degrees correspond to characteristic Al peaks, as highlighted with the inverse triangle symbols (▼). These minor peaks suggest the presence of the Al phase, indicating that the dealloying process is not fully complete in 3 minutes [17]. Although the 3-minute dealloyed NP-Zn requires additional leaching, keeping this sample in the corroding medium for two more minutes (i.e., from 3 to 5 minutes) leads to the formation of ZnO, as shown by the orange XRD pattern of the 5-minute dealloyed sample in Fig. 5.3, which shows that the product powder is mostly metallic Zn after dealloying for 5 minutes; however, minor peaks of ZnO highlighted with square signs (■) are observed, suggesting

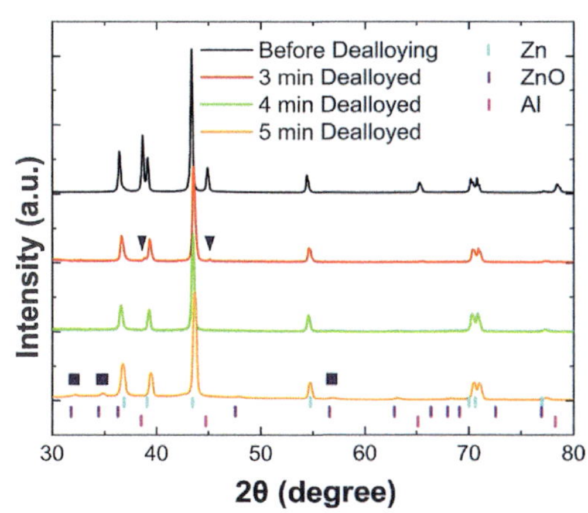

Figure 5.3 XRD patterns of Zn-Al parent alloy before dealloying and NP-Zn after dealloying for 3 minutes (red), 4 minutes (green), and 5 minutes (orange). NP-Zn dealloyed for 4 minutes exhibits pure Zn patterns lacking Al and ZnO reflections. NP-Zn dealloyed for 3 minutes still contains residual Al, as highlighted with the inverse triangles (▼), and Zn starts to oxidize after 5 minutes of dealloying, forming ZnO on the surface, as highlighted with the square (■).

that Zn has started to oxidize when Al removal is completed. The optimal dealloying time for the Zn-Al parent alloy powder is determined to be 4 minutes, as shown in the green XRD pattern in Fig. 5.3, where it can be seen that NP-Zn powder leached for 4 minutes has neither Al nor ZnO.

In addition to the XRD analysis, EDS (Energy-dispersive X-ray spectroscopy) and ICP-OES (Inductively Coupled Plasma - Optical Emission Spectrometry) confirmed the removal of Al. EDS was performed on particle cross-sections to determine the composition using FIB/SEM. As shown in Table 5.1, after 4 minutes of selective etching, 99% of the product powder consists of Zn (from EDS), indicating that nearly all Al has been successfully removed in the NaOH solution. EDS also suggests that the Zn-Al parent alloy (nominal composition $Zn_{40}Al_{60}$) consists of 51 at. % Al and 49 at. % Zn. Since our EDS data gives the chemical composition within a powder particle, we attribute this 9 at. % deviation from the nominal composition to the fluctuation in composition among different powder particles. In other words, this deviation would be less if we take the average chemical composition of many powder particles (interparticle composition) instead of within one single particle (intra-particle composition). This motivated us to run ICP to investigate the sample composition more accurately since ICP can give the average composition of a large amount of powder dissolved in acid. The results from Table 5.1 demonstrate that the overall chemical composition from ICP of Al and Zn in the parent Zn-Al alloy is 65.9 at. % and 34.1 at. %, respectively, which is closer to the nominal composition of $Zn_{40}Al_{60}$ at. % than the EDS data. ICP data also suggests that 4 minutes of dealloying can dissolve most of the Al phase, and the etched powder only has 6 at. % of Al remaining (Table 5.1). Both EDS and ICP characterization confirm the successful removal of Al from the Zn-Al alloy quantitatively.

Table 5.1 EDS and ICP-based elemental analysis of Zn-Al powder before and after dealloying.

Dealloy time	Al (at%)	Zn (at%)
ICP		
Before	65.9	34.1
After	6.0	94.0
EDS		
Before	51.3	48.7
After	0.7	99.3

The pristine Zn-Al powder is further investigated using EDS mapping. Fig. 5.4a and b illustrate the coexistence of two phases in Zn-Al powder, while the EDS mappings in Fig. 5.4c—e confirm that the purple regions correspond to Zn and the orange areas to Al; it signifies the phase separation of Al and Zn phase in the parent alloy powder. This interconnected structure originates from the immiscibility gap of the Zn-Al material system [26].

As the Zn-Al melt cools during gas atomization, the alloy experiences spinodal decomposition of the Al and Zn phases. Spinodal decomposition is a process where a homogeneous mixture spontaneously separates into different phases upon becoming thermodynamically unstable, leading to compositional fluctuations [27]. It should also be noted that there are small

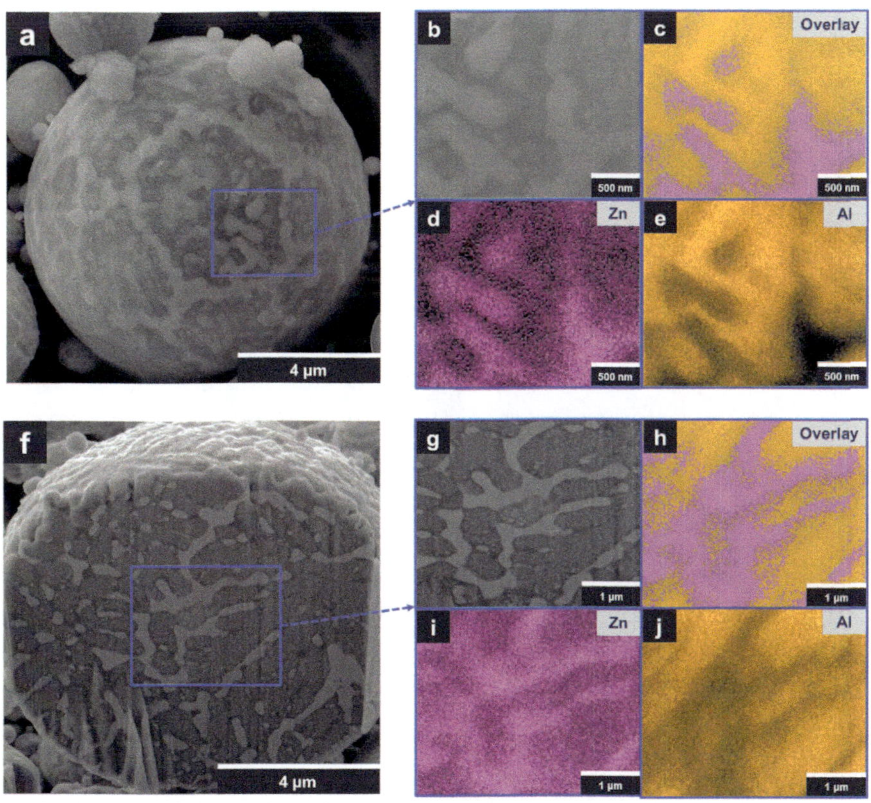

Figure 5.4 SEM images with corresponding EDS maps of (a—e) the surface and (f—j) the cross section of the Zn-Al parent alloy. (a,f) Low-magnification SEM images illustrate the separation of two phases. (b,g) High-magnification SEM images demonstrate small particles of Zn within the Al phase. The corresponding EDS maps show (c,h) Zn (pink) and Al (orange) overlay, (i) Zn in pink, and (j) Al in orange.

white precipitations in the dark Al regions in Fig. 5.4f and g (see also the high magnification images used in the graphical abstract). According to the EDS mapping analysis, the white phase corresponds to Zn, indicating that these small precipitates are Zn within the Al matrix. Such a hierarchical structure of Zn creates a unique morphology once the powder is dealloyed.

After dealloying, FIB/SEM images of NP-Zn reveal a hierarchical nanoporous structure featuring large ligaments/pores of approximately 100–200 nm, Fig. 5.5b, and smaller ligaments/pores of roughly 50–60 nm, Fig. 5.5g in Fig. 5.5. The large nanoporous structures are the remaining Zn phase from the phase-separated Zn-Al parent alloy after the

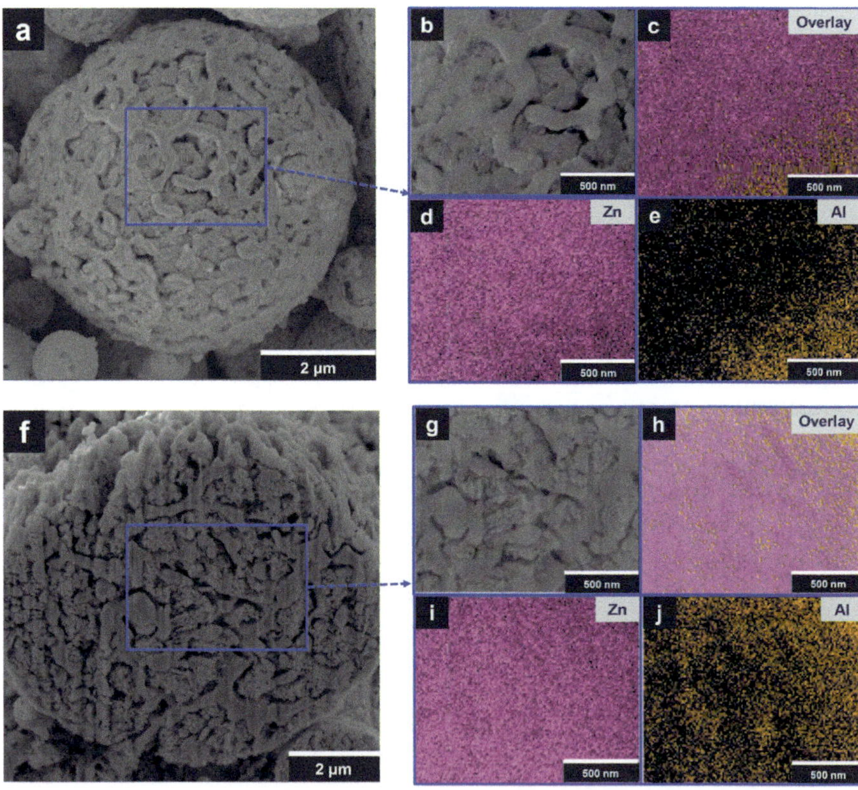

Figure 5.5 SEM images with corresponding EDS maps of (a–e) the surface and (f–j) the cross section of NP-Zn with hierarchical nanoporous structure. (a,f) Low-magnification and (b,g) high-magnification SEM images showing the large ligament/pore features in the 100–200 nm range and the small ligament/pore features in the 50–60 nm range. The corresponding EDS maps show (c,h) Zn (pink) and Al (orange) overlay, (i) Zn in pink, and (j) Al in orange, indicating that NP-Zn consists primarily of Zn.

Al phase has been removed. The Zn could sustain its morphology while Al is selectively etched away because the two phases are already separated before the dealloying. The small nanoporous structures originate from the Zn nanoparticles trapped in the Al matrix during alloy solidification (see the high-magnification images used for the graphical abstract). When the Al matrix is etched away during dealloying, these Zn nanoparticles will agglomerate with each other to form the nanoscale Zn features observed [28]. Both EDS mappings on the powder surface (Fig. 5.5c) and the cross section (Fig. 5.5h) confirm the almost complete removal of Al atoms in NP-Zn. Some signals indicate Al in the map in Fig. 5.5e and j. Still, the EDS spectra show that the intensity corresponding to Al is almost negligible.

It is emphasized that the dealloying time for our NP-Zn powder is significantly much shorter than the previous study in Chapter 4 with bulk NP-Zn [24]. The bulk alloy took 3 days to remove Al, but the present study shows that Zn-Al powder can be dealloyed in 4 minutes. The dealloying time is reduced because the micrometer-sized Zn-Al parent alloy powder has a higher surface area and phase-separated Al and Zn phases. The micrometer-sized powder offers more surface area for reacting with the alkaline electrolyte than the centimeter-sized alloy chunk, enhancing the reaction kinetics for Al dissolution. Furthermore, unlike the monolithic single-phase bulk alloy used in previous work, our Zn-Al powder contains two distinct regions: Al-rich and Zn-rich. With these phases already separated before the dealloying process, Al is more readily removed and dissolved.

5.5 Enhanced reactivity of nanoporous Zn powder fuel with water as the oxidizer

Chapter 4 (Section 4.6) demonstrates the enhanced reactivity of the dealloyed monolithic bulk NP-Zn as a wet metal fuel, using H_2O as the oxidizer. Similarly, in this section, we aim to show that dealloyed NP-Zn in powder form is even more reactive as a wet metal fuel than monolithic bulk NP-Zn. In doing so, we use the setup in Fig. 5.6a to show that the fabricated NP-Zn in the powder form significantly enhances the kinetics of the reaction between Zn and water to produce hydrogen and zinc oxide (ZnO) solid by-product through Eq. (1.12), just like we did in Chapter 4:

$$Zn(s) + H_2O(l) \rightarrow ZnO(s) + H_2(g) + \text{heat} \tag{5.4}$$

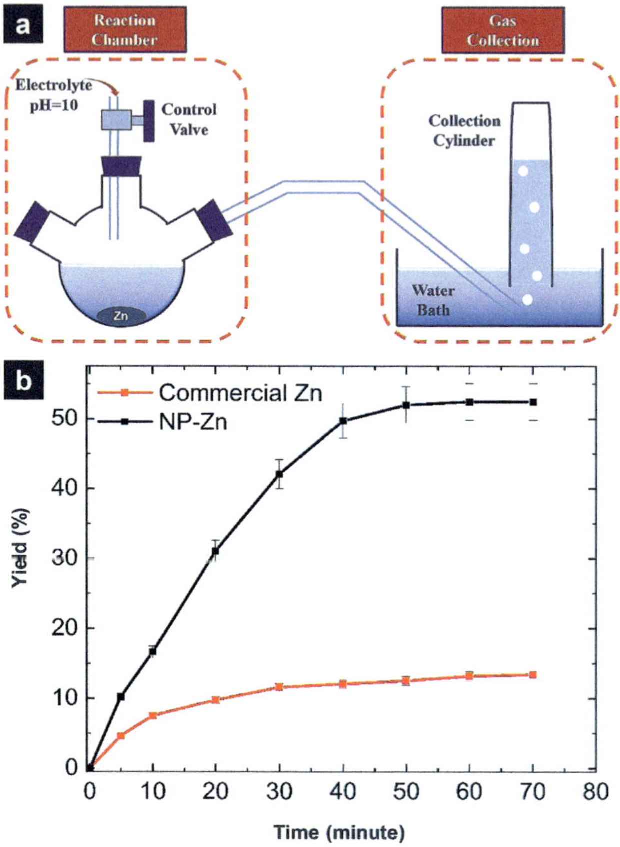

Figure 5.6 (a) Schematic illustration of water displacement setup used to measure the hydrogen yield. (b) The hydrogen yield of commercial Zn (red) and NP-Zn (black) as a function of the reaction time. NP-Zn significantly improves the metal-water reaction kinetics due to its nanoporous structure.

In particular, we compare the yield of hydrogen produced from our NP-Zn powder (black curve Fig. 5.6b) with that of commercial Zn powder without porosity (red curve Fig. 5.6b). After 1 hour of reaction, the hydrogen yield for the commercial Zn powder is ~13%. The hydrogen generation rate significantly slows down after 30 minutes due to forming a passive layer of ZnO (dark blue shell in Fig. 5.7a) around the starting Zn particle. This ZnO shell prevents water from directly contacting metallic Zn underneath to react further, as explained in Chapter 4 [24,29,30]. On the other hand, a yield of ≈52% is achieved on our NP-Zn within the first 40–50 minutes.

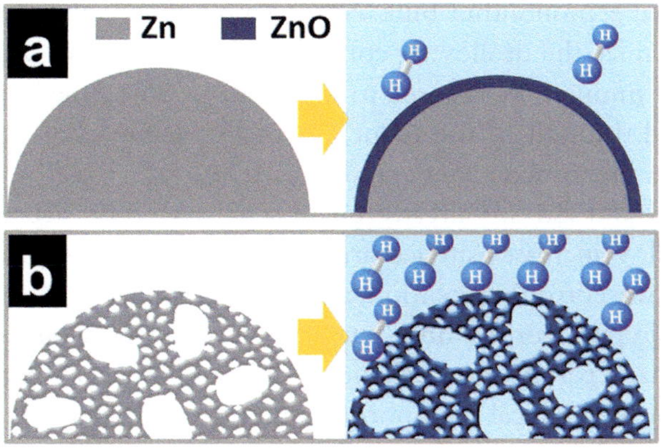

Figure 5.7 Schematics of Zn metal-water reaction producing gaseous H_2 and solid ZnO by-product. (a) Once bulk Zn reacts with water, a layer of ZnO is formed, making the metallic Zn core inaccessible. (b) Zn in NP-Zn can fully react with water due to the porous structure and the high surface-area-to-volume ratio, overcoming the issue of the water-blocking oxide layer [32].

This substantial improvement in hydrogen evolution kinetics is attributed to the high surface area and hierarchical porous structure of the NP-Zn powder. The oxidation of Zn is a near-surface reaction, and the extensive surface area provided by the numerous tiny pores, as shown in Fig. 5.5f and g, introduces more reaction sites than commercial Zn dust. While these small pores enhance the metal-electrolyte interface, the larger pores serve as channels that allow the evolved hydrogen gas to escape. This pathway is advantageous for hydrogen generation because the produced hydrogen gas forms bubbles that can obstruct the surface of the metallic Zn, impeding further oxidation [31]. The hierarchical porous structure facilitates the removal of these bubbles, ensuring a continuous and efficient reaction process. The large pores also effectively improve the transport of electrolytes, which can induce more reactions at the inner surface of NP-Zn. We expect further optimization of the dealloying process, and the post-dealloying process can increase the hydrogen production capacity in this material system.

5.6 Conclusions

Chapter 5 demonstrates the rapid fabrication of NP-Zn by free corrosion dealloying of commercially available micrometer-sized $Zn_{40}Al_{60}$ parent alloy produced by gas atomization. Prior to this work, the selective etching process to make monolithic bulk NP-Zn typically requires several days. Here we optimize the dealloying conditions for making NP-Zn in powder form in just 4 minutes, as opposed to several days. Our expeditious dealloying is enabled through (i) the reduction of the parent alloy size from the centimeter to micrometer range and (ii) the use of Zn-Al parent alloy in which the Al and Zn are already phase-separated. The size reduction in the parent alloy increases surface area, enabling more sacrificial elements to react with the corroding solution, accelerating in that way the free corrosion dealloying rate. Furthermore, the separation of the Al and Zn phases in the gas-atomized parent alloy enhances Al dissolution. The dealloyed NP-Zn exhibits a hierarchical nanoporous structure with small ligament/pore structures around 50−60 nm and large ligament/pore structures around 100−200 nm. The large ligament/pore structures correspond to the Zn phase from the phase-separate Zn-Al alloy, after the Al phase has been etched away. The small ligament/pore structures result from the agglomeration of Zn nanoparticles embedded in the Al phase, when this Al phase is etched away during dealloying. We also studied how the nanoporous morphology influences the Zn-water reaction, which produces hydrogen and ZnO as the solid by-product. We found that the dealloyed NP-Zn powder can produce hydrogen from water with a yield of ≈52% within 40 minutes, while the yield of the conventional dense Zn powder is only ≈13%, demonstrating significant improvement in the reaction kinetics. The hierarchical structure and porous surface facilitate efficient diffusion of the produced hydrogen gas and provide a more accessible metallic surface to react with the solution, respectively. Hydrogen production with NP-Zn through metal-water reaction is promising for large-scale applications.

Acknowledgments

Chapter 5: Part of the content of this chapter has been published by J. Kubota, J. Liu, and Eric Detsi in ACS Applied Materials & Interfaces 16 (2024), entitled, *Rapid Synthesis of Nanoporous Zn Powder by Selective Etching of Al from Micrometer-Sized Zn-Al Powder Particles Produced by Gas-Atomization and Its Application in Hydrogen Generation*. The authors are thankful to Jo Kubota and Jiaxin Liu.

References

[1] J. Fu, S.S. Welborn, E. Detsi, Dealloyed air- and water-sensitive nanoporous metals and metalloids for emerging energy applications, ACS Appl. Energy Mater. 5 (6) (2022) 6516−6544.

[2] J. Snyder, I. McCue, K. Livi, J. Erlebacher, Structure/processing/properties relationships in nanoporous nanoparticles as applied to catalysis of the cathodic oxygen reduction reaction, J. Am. Chem. Soc. 134 (20) (2012) 8633−8645.

[3] C. Zhao, T. Wada, V. De Andrade, D. Gürsoy, H. Kato, Y.-c.K. Chen-Wiegart, Imaging of 3D morphological evolution of nanoporous silicon anode in lithium ion battery by X-ray nano-tomography, Nano Energy 52 (2018) 381–390.
[4] Y. An, H. Fei, G. Zeng, L. Ci, S. Xiong, J. Feng, Y. Qian, Green, scalable, and controllable fabrication of nanoporous silicon from commercial alloy precursors for high-energy lithium-ion batteries, ACS Nano 12 (5) (2018) 4993–5002.
[5] H.-J. Jin, J. Weissmüller, Bulk nanoporous metal for actuation, Adv. Eng. Mater. 12 (8) (2010) 714–723.
[6] J. Biener, A. Wittstock, L.A. Zepeda-Ruiz, M.M. Biener, V. Zielasek, D. Kramer, R.N. Viswanath, J. Weissmüller, M. Bäumer, A.V. Hamza, Surface-chemistry-driven actuation in nanoporous gold, Nat. Mater. 8 (1) (2009) 47–51.
[7] A.N. Koya, X. Zhu, N. Ohannesian, A.A. Yanik, A. Alabastri, R. Proietti Zaccaria, R. Krahne, W.-C. Shih, D. Garoli, Nanoporous metals: from plasmonic properties to applications in enhanced spectroscopy and photocatalysis, ACS Nano 15 (4) (2021) 6038–6060.
[8] D. Garoli, E. Calandrini, G. Giovannini, A. Hubarevich, V. Caligiuri, F. De Angelis, Nanoporous gold metamaterials for high sensitivity plasmonic sensing, Nanosc. Horiz. 4 (5) (2019) 1153–1157.
[9] Y. Ding, M. Chen, Nanoporous metals for catalytic and optical applications, MRS Bull. 34 (8) (2009) 569–576.
[10] I.V. Okulov, J. Weissmüller, J. Markmann, Dealloying-based interpenetrating-phase nanocomposites matching the elastic behavior of human bone, Sci. Rep. 7 (1) (2017) 20.
[11] T.J. Balk, C. Eberl, Y. Sun, K.J. Hemker, D.S. Gianola, Tensile and compressive microspecimen testing of bulk nanoporous gold, JOM 61 (12) (2009) 26–31.
[12] Y. Ding, Y.J. Kim, J. Erlebacher, Nanoporous gold leaf: "Ancient Technology"/advanced material, Adv. Mater. 16 (21) (2004) 1897–1900.
[13] C. Coaty, H. Zhou, H. Liu, P. Liu, A scalable synthesis pathway to nanoporous metal structures, ACS Nano 12 (1) (2018) 432–440.
[14] M.C. Dixon, T.A. Daniel, M. Hieda, D.M. Smilgies, M.H.W. Chan, D.L. Allara, Preparation, structure, and optical properties of nanoporous gold thin films, Langmuir 23 (5) (2007) 2414–2422.
[15] D.V. Pugh, A. Dursun, S.G. Corcoran, Formation of nanoporous platinum by selective dissolution of Cu from Cu0.75Pt0.25, J. Mater. Res. 18 (1) (2003) 216–221.
[16] X. Wang, W. Wang, Z. Qi, C. Zhao, H. Ji, Z. Zhang, Electrochemical catalytic activities of nanoporous palladium rods for methanol electro-oxidation, J. Power Sources 195 (19) (2010) 6740–6747.
[17] S. Niauzorau, N. Kublik, E. Dasinor, A. Hasib, A. Sharstniou, B. Azeredo, Casting of high surface area electrodes enabled by low-temperature welding of copper nanoporous powders and nanoparticles hybrid feedstocks, Appl. Mater. Today 32 (2023) 101802.
[18] T. Fujita, L.-H. Qian, K. Inoke, J. Erlebacher, M.-W. Chen, Three-dimensional morphology of nanoporous gold, Appl. Phys. Lett. 92 (25) (2008).
[19] W. Moschkowitsch, O. Lori, L. Elbaz, Recent progress and viability of PGM-free catalysts for hydrogen evolution reaction and hydrogen oxidation reaction, ACS Catal. 12 (2) (2022) 1082–1089.
[20] K.-L. Yan, J.-F. Qin, Z.-Z. Liu, B. Dong, J.-Q. Chi, W.-K. Gao, J.-H. Lin, Y.-M. Chai, C.-G. Liu, Organic-inorganic hybrids-directed ternary NiFeMoS anemone-like nanorods with scaly surface supported on nickel foam for efficient overall water splitting, Chem. Eng. J. 334 (2018) 922–931.

[21] Y. Ren, Y. Yang, M. Wei, Recent advances on heterogeneous non-noble metal catalysts toward selective hydrogenation reactions, ACS Catal. 13 (13) (2023) 8902–8924.
[22] L. Lu, Nanoporous noble metal-based alloys: a review on synthesis and applications to electrocatalysis and electrochemical sensing, Microchim. Acta 186 (9) (2019) 664.
[23] R. Song, J. Han, M. Okugawa, R. Belosludov, T. Wada, J. Jiang, D. Wei, A. Kudo, Y. Tian, M. Chen, H. Kato, Ultrafine nanoporous intermetallic catalysts by high-temperature liquid metal dealloying for electrochemical hydrogen production, Nat. Commun. 13 (1) (2022) 5157.
[24] J. Fu, Z. Deng, T. Lee, J.S. Corsi, Z. Wang, D. Zhang, E. Detsi, pH-controlled dealloying route to hierarchical bulk nanoporous Zn derived from metastable alloy for hydrogen generation by hydrolysis of Zn in neutral water, ACS Appl. Energy Mater. 1 (7) (2018) 3198–3205.
[25] E. McCafferty, Thermodynamics of corrosion: Pourbaix diagrams, in: E. McCafferty (Ed.), Introduction to Corrosion Science, Springer New York, New York, NY, 2010, pp. 95–117.
[26] J.L. Murray, The Al–Zn (Aluminum-Zinc) system, Bull. Alloy Phase Diagr. 4 (1) (1983) 55–73.
[27] X. Zhou, R. Darvishi Kamachali, B.L. Boyce, B.G. Clark, D. Raabe, G.B. Thompson, Spinodal decomposition in nanocrystalline alloys, Acta Mater. 215 (2021) 117054.
[28] J. Erlebacher, M.J. Aziz, A. Karma, N. Dimitrov, K. Sieradzki, Evolution of nanoporosity in dealloying, Nature 410 (6827) (2001) 150–153.
[29] V. Rosenband, A. Gany, Application of activated aluminum powder for generation of hydrogen from water, Int. J. Hydrogen Energy 35 (20) (2010) 10898–10904.
[30] X.-n. Huang, C.-j. Lv, Y.-x. Huang, S. Liu, C. Wang, D. Chen, Effects of amalgam on hydrogen generation by hydrolysis of aluminum with water, Int. J. Hydrogen Energy 36 (23) (2011) 15119–15124.
[31] L.-P. Han, R.-Q. Yao, W.-B. Wan, H. Shi, Z. Wen, X.-Y. Lang, Q. Jiang, Hierarchical nanoporous intermetallic compounds with self-grown transition-metal hydroxides as bifunctional catalysts for the alkaline hydrogen evolution reaction, J. Mater. Chem. A 7 (45) (2019) 25925–25931.
[32] J.S. Corsi, Fabrication and Study of Microstructure, Solid-Electrolyte Interphase, and Phase Evolution of Nanoporous Metals Used as Li-Ion Battery Alloy Anodes, UPenn, 2021. https://repository.upenn.edu/entities/publication/39e8d876-f749-4140-916b-6728148bcf63.

Further reading

[1] J.S. Corsi, J. Fu, Z. Wang, T. Lee, A.K. Ng, E. Detsi, Hierarchical bulk nanoporous aluminum for on-site generation of hydrogen by hydrolysis in pure water and combustion of solid fuels, ACS Sustain. Chem. Eng. 7 (13) (2019) 11194–11204.

CHAPTER SIX

Monolithic bulk nanoporous aluminum by air-free electrolytic dealloying with recovery of sacrificial materials

6.1 Overview

As aforementioned in Chapter 4, the specific format of nanoporous metal fuels, such as a monolithic piece of nanoporous metal, and nanoporous metal in powder form, is critical for their *processability* and *reactivity*. This issue was addressed in particular by comparing monolithic bulk nanoporous Zn (Chapter 4) and nanoporous Zn in powder form (Chapter 5) in terms of their *processability* (i.e., fabricating monolithic bulk nanoporous Zn vs. fabricating nanoporous Zn in powder form) and *reactivity* (reacting monolithic nanoporous Zn vs. nanoporous Zn in powder form with H_2O to produce hydrogen and Zn oxide as the solid by-product). In the present chapter, another example is examined based on monolithic bulk nanoporous Al (Chapter 6) and nanoporous Al in powder form (Chapter 7). In doing so, a comparison between monolithic bulk nanoporous Al (Chapter 6) and nanoporous Al in powder form (Chapter 7) will be made in terms of their *processability* (i.e., fabricating monolithic bulk nanoporous Al vs. fabricating nanoporous Al in powder form) and *reactivity* (reacting monolithic nanoporous Al vs. nanoporous Al in powder form with H_2O to produce hydrogen and Al hydroxide as the solid by-product).

Here, first the fabrication of monolithic bulk nanoporous Al is addressed made from a monolithic bulk Al—Mg parent alloy, using a novel and sustainable air-free electrolytic dealloying process in which the sacrificial Mg is recovered in real-time during electrolytic dealloying. Next, the reactivity of the fabricated nanoporous Al with neutral water to produce hydrogen gas and Al hydroxide solid by-product is investigated. Since Al hydroxide formed during the reaction of Al with H_2O is self-limiting (as we discussed in previous chapters), we hypothesize that there exists a critical feature size of Al below which the Al—H_2O reaction will fully proceed without self-

Metals as Clean Fuels
ISBN: 978-0-443-13537-8
https://doi.org/10.1016/B978-0-443-13537-8.00009-9

limitation. We further hypothesize that this critical feature size can be approximated with the minimum thickness of the passivating Al hydroxide layer formed during the reaction of Al with H_2O, which has been reported to be around ~3.5 nm [1].

Based on our hypotheses, nanostructured Al with a feature size around ~7 nm (i.e., twice the thickness of the Al hydroxide layer) should fully react with H_2O to produce Al hydroxide and hydrogen with a yield of 100%. Consequently, we should make nanoporous Al with ligament size around 7 nm or less to achieve full reactivity. However, making such ultrafine nanoscale Al will be challenging. Therefore, we set a goal to make nanoscale Al with a feature size in the range of 10—20 nm. While nanoscale Al with a feature size of around 10—20 nm may not fully react with neutral H_2O to produce hydrogen and Al hydroxide, the enhanced reactivity is expected to be significant. So, we ultimately make hierarchical nanoporous Al consisting of macroscopic and mesoscopic ligament/pore structures. The mesoscopic ligaments in the range of 10—20 nm actively react with water to produce hydrogen with a yield of ~52%—90%, without incorporation of any catalyst or reaction promoter in the Al—H_2O system. The macroscopic pores, in the range of 0.5-1 um, facilitate the flow of H_2O through the bulk of Al, allowing the mesoscopic ligaments to react with water, as discussed in the previous chapters for the reaction between hierarchical nanoporous Zn with H_2O. Thus, the present chapter will cover the following topics: (1) Fabricating oxide-free hierarchical bulk nanoporous Al with ultrafine ligaments in the range of 10—20 nm by electrolytic removal of sacrificial Mg from Al—Mg parent alloys with near-eutectic composition; (2) Demonstrating the recovery of the sacrificial Mg simultaneously during the electrolytic leaching process. Recovered Mg can be reused to make new Al—Mg parent alloys; (3) Wet metal fuel oxidation, during which we use the fabricated hierarchical nanoporous Al with water to produce hydrogen, heat, and Al hydroxide. Hydrogen gas is produced with a yield in the range of ~52% —90%; Also, not shown here, dry metal fuel oxidation is studied during which we qualitatively found that the fabricated bulk nanoporous Al can really ignite in air.

6.2 Experimental methods

Al—Mg parent alloys with near-eutectic composition were made by homogeneously melting pure Al and pure Mg at 750°C in a graphite crucible using a quartz tube under argon flow in combination with a tube furnace [2]. The nominal composition of the parent alloy was $Al_{30}Mg_{70}$ at. %. Nanoporous Al was fabricated by selective electrolytic removal of Mg from the above-mentioned Al–Mg parent alloys using a Bio-Logic VMP-300 potentiostat/galvanostat [3]. A JEOL 7500F scanning electron microscope (SEM) equipped with energy dispersive X-ray spectroscopy (EDS) capability was used to characterize the microstructure and elemental composition of the fabricated Al–Mg parent alloy and corresponding dealloyed nanoporous Al. Inductively coupled plasma optical emission spectroscopy (Spectro Genesis ICP-OES device) was also used to further investigate the bulk chemical compositions of these materials. The X-ray diffraction (XRD) data of these materials was taken with a Rigaku D/Max-B X-ray diffractometer with Bragg – Brentano parafocusing geometry operating with Cu Kα1 line ($\lambda = 1.5405$ Å) and a horizontal goniometer collecting at an angle range of 10–95 degrees with 0.05 degrees step size and 2 degrees/min scanning speed, with a conventional copper target X-ray tube set to 40 kV and 30 mA. A customized gas chromatograph system (GC, Scion 456) equipped with a pulsed discharge detector (PDD) with 5 Å molecular sieve-packed column, and a thermal conductivity detector (TCD) was used to measure the amount and rate of hydrogen produced during the reaction of nanoporous Al with water [4]. X-ray photoelectron spectroscopy (XPS) was carried out utilizing a PHI VeraProbe 5000 instrument equipped with a monochromated Al Kα radiation source in order to investigate the chemical composition of the solid product formed during the reaction of Al with water.

6.3 Characterization of the $Al_{30}Mg_{70}$ parent alloy

The Al–Mg parent alloy with near-eutectic composition was used for the fabrication of hierarchical nanoporous Al. In this section, we present the characterization of this parent alloy, which was made by melting pure Al and pure Mg (see Experimental methods in Section 6.2 above). Two microstructural phases were present in the fabricated near-eutectic Al–Mg alloy [5], as expected from the Al–Mg equilibrium phase diagram. These phases can be distinguished from the low-magnification scanning electron micrograph in Fig. 6.1a, where the dark area corresponds to (Mg) solid solution [5], and the light area represents β-$Al_{12}Mg_{17}$ intermetallic (the alloy surface was pretreated by etching to enhance the phase contrast) [5]. Note that it is possible to identify these two phases from low-magnification imaging because the alloy surface was pretreated with acid to initiate the corrosion

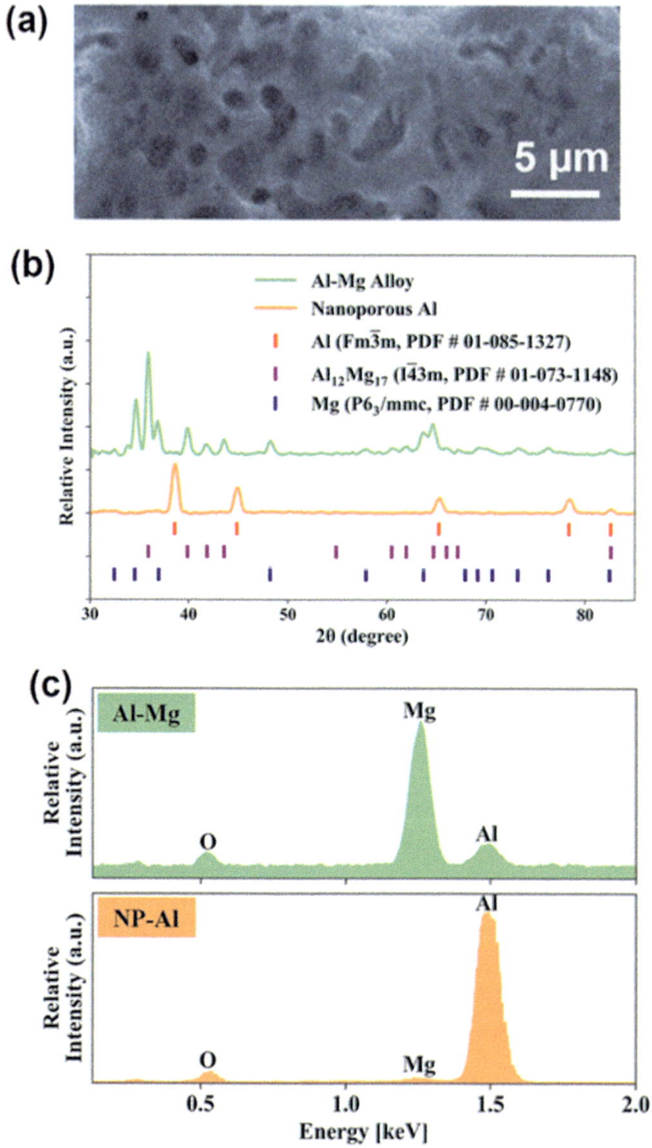

Figure 6.1 (a) SEM of the starting Al—Mg parent alloy with the surface pretreated to enhance the contrast between the (Mg) phase (dark) and β-$Al_{12}Mg_{17}$ phase (light). (b) XRD data for the Al—Mg parent alloy before (green) and after (orange) dealloying, together with crystallography data of Al (red), $Al_{12}Mg_{17}$ (purple), and Mg (blue). (c) EDS data of the Al—Mg parent alloy before (green) and after (orange) dealloying.

process; since the (Mg) phase dissolves faster than the β-$Al_{12}Mg_{17}$ intermetallic phase after surface treatment, the two phases can be dissociated from low-magnification electron micrographs. The fabricated parent alloy was further characterized by XRD to confirm the presence of these two phases. The green curve in Fig. 6.1b represents the typical XRD pattern of the Al–Mg parent alloy. All the diffraction peaks from this pattern match the hexagonal Mg metal ($P6_3/mmc$) with PDF reference card number 00-004-0770, as depicted by the blue stick pattern in Fig. 6.1b; and the cubic $Al_{12}Mg_{17}$ (I-43m) with PDF reference card number 01-073-1148 as depicted by the purple stick pattern in Fig. 6.1b. Finally, the real chemical composition of the fabricated Al–Mg parent alloy with nominal composition $Al_{30}Mg_{70}$ at. % was investigated by EDS and ICP. The green EDS spectrum in Fig. 6.1c shows the typical elemental composition of this alloy. Al and Mg signals are observed as expected. Quantitatively, EDS data suggest that the alloy composition is $Al_{24}Mg_{76}$ at. %, which corresponds to a hypoeutectic composition [5]. The small oxygen signal observed on the EDS spectrum of the parent alloy can be attributed to the native oxide formed upon the reaction of Mg and Al with air. Because the interaction volume sampled by the electron beam is relatively shallow in EDS, the chemical composition obtained from EDS is more a representation of the near-surface composition. Therefore, ICP was used to determine the bulk composition of our alloy [4], which was found to be ~$Al_{37}Mg_{63}$ at. % (i.e., hypereutectic composition [5]). Both EDS and ICP data are summarized in Table 6.1 below.

Table 6.1 Energy-dispersive X-ray spectroscopy (EDS) and Inductively Coupled Plasma - Optical Emission Spectrometry (ICP-OES) elemental composition analysis of the Al–Mg parent alloy before and after dealloying.

		Al (at. %)	Mg (at. %)
Before dealloying	ICP[a]	36.6	63.4
	EDS	24	76
After dealloying	ICP[a]	94.3	5.7
	EDS	98	2

[a]The samples before and after dealloying were dissolved in a mixture of HNO_3 and HCl for ICP analysis.

6.4 Synthesis and characterization of monolithic bulk nanoporous Al by air-free electrolytic dealloying with recovery of sacrificial Mg

In the following, the fabrication of hierarchical nanoporous Al with ultrafine ligament/pore structures will be demonstrated using a novel air-free electrolytic dealloying route in nonaqueous electrolytes [6]. As discussed in previous chapters, although a wide range of nanoporous metals [7–22] and nanoporous metalloids [23–26] have been fabricated using the conventional chemical and electrolytic dealloying route in aqueous electrolytes, these aqueous electrolytes are not suitable for the fabrication of reactive nanoporous metals [4], such as nanoporous Al, because ionic species such as H^+ and OH^- present in aqueous electrolytes will spontaneously react with Al. In a first attempt to overcome this limitation, it was shown feasible to synthesize ultrafine nanoporous Al with ligament size in the range of ~10 nm by galvanic replacement in nonaqueous electrolytes [27]. Detrimental side reactions occur during this galvanic replacement dealloying reaction, including the co-deposition (plating) of Al concurrently during selective leaching of Mg, which results in severe structural coarsening (i.e., large ligaments). To avoid this coarsening issue, a two-step synthesis route involved the formation of ultrafine nanoporous aluminum hydride composite structures by galvanic replacement, followed by heat-induced hydrogen desorption from these hydrides to obtain nanoporous Al with ligament size in the range of ~10 nm [27]. Recently, Jin and coworkers also exploited the galvanic replacement method to fabricate nanoporous Al with ligament size in the range of 200 nm [28]. To overcome the limitation of the galvanic replacement approach (i.e., co-deposition of Al and structural coarsening), we use a novel air-free nonaqueous electrolytic dealloying method [6]. In this approach, the reduced working ion in the nonaqueous electrolyte (i.e., Mg^{2+}) is the same nature as the sacrificial element (i.e., Mg) in the Al–Mg parent alloy. In such a configuration, it is possible to electrochemically remove Mg from the starting Al–Mg alloy and plate it onto the counter electrode, resulting in the simultaneous recovery of the sacrificial element during selective leaching. Such a recovery of the sacrificial material is not possible during dealloying by galvanic replacement, or during conventional dealloying in aqueous electrolyte where the sacrificial element is not the same nature as the reduced working ion.

In our work, the electrolytic cell configuration consists of a 3-neck round bottom flask with the Al–Mg parent alloy used as the working electrode, a pure Mg foil used as the counter and reference electrodes, and 0.4 M all-phenyl complex (APC) electrolyte used as Mg-ion conducting electrolyte. The APC electrolyte was prepared following the procedure from previous work [3]. The selective leaching process was carried out under an inert atmosphere in an argon-filled MBraun glovebox with H_2O and O_2 content below 0.1 ppm [3]. All the experiments were performed at room temperature under ambient pressure in the glovebox. Mg was selectively removed from the Al—Mg parent alloy using electrolytic dealloying under potentiostatic control [29,30]. During potentiostatic controlled electrolytic dealloying, a constant voltage is maintained between the working and reference electrodes in order to dissolve Mg, and the corresponding Mg dissolution current between the working and counter electrodes is measured [30]. Fig. 6.2a shows the typical electrolytic cell configuration before dealloying. A starting bulk piece of Al—Mg alloy with dimensions in the centimeter range was used as the working electrode, and an Mg foil of similar size was used as both the counter and reference electrodes. Fig. 6.2b shows the cell configuration after dealloying, as well as the fabricated bulk piece of nanoporous Al. The dealloyed nanoporous Al was broken into two pieces to demonstrate its brittleness, being a common property of dealloyed nanoporous metals [31]. The Mg foil counter electrode after dealloying is also shown in Fig. 6.2b. It can be seen that the surface of this Mg foil, which was initially smooth and dull before dealloying (see Fig. 6.2a) has become very rough and shiny after dealloying. This is a result of the Mg plating process, which occurs during selective leaching. To verify the chemical nature of the plated Mg, it was scraped from the counter electrode, collected in powder form, and melted into a monolithic piece for characterization by XRD (see Fig. 6.2c) and EDS. Both characterization methods confirm the formation of high-purity Mg.

The green line in Fig. 6.3 represents a typical current profile as a function of time associated with Mg removal from the Al—Mg working electrode and Mg plating onto the Mg foil counter electrode at a constant potential of 450 mV versus Mg/Mg^{2+}. By integrating this current with respect to time, the total (cumulative) electric charge transferred during dealloying can be calculated. Faraday's laws of electrolysis were then used to convert this cumulative charge into the mass of Mg removed from the starting Al—Mg working electrode [29] and plated on the Mg counter electrode during dealloying as shown by the orange curve in Fig. 6.3 [29]. The electrolytic dealloying process was finished when the mass of Mg from the starting Al—

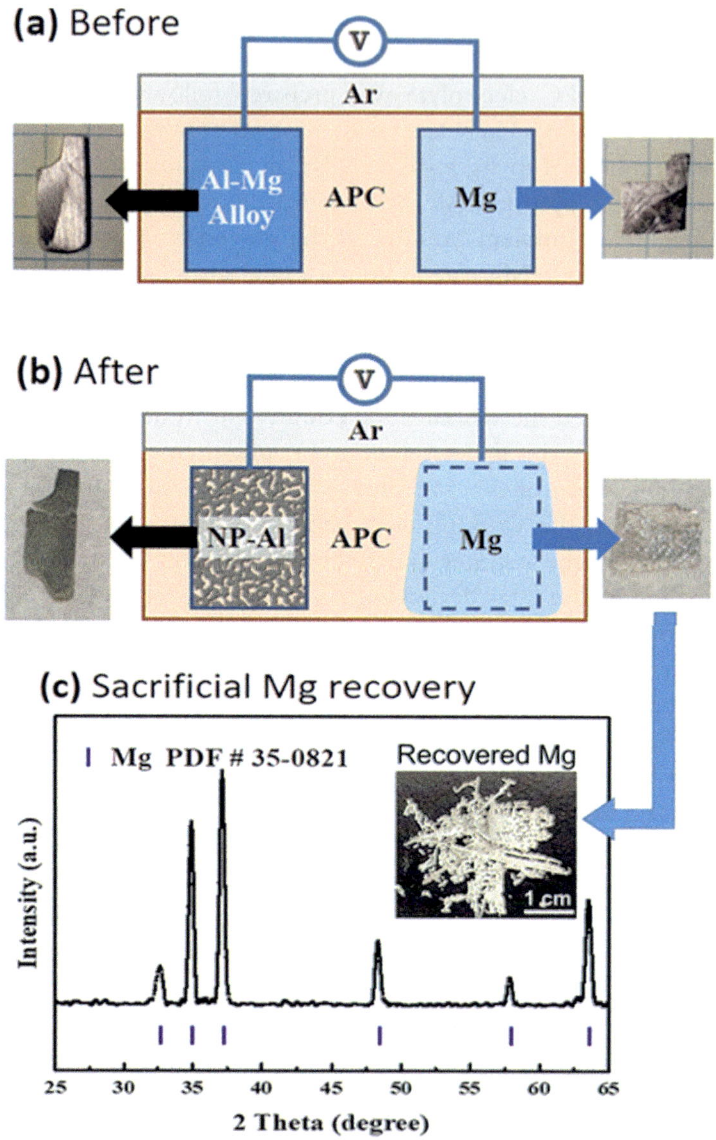

Figure 6.2 (a) Schematic of the electrolytic cell setup before dealloying. The cell contains an Al–Mg parent alloy used as the working electrode and Mg foil used as counter and reference electrodes in APC electrolyte. The cell operates in an argon environment. (b) Corresponding schematic of the electrolytic cell after dealloying. The starting Al–Mg working electrode is converted into nanoporous Al. Meanwhile, Mg removed from this parent alloy is plated onto the Mg foil counter electrode. (c) XRD data of stripped Mg recovered on the counter electrode after dealloying.

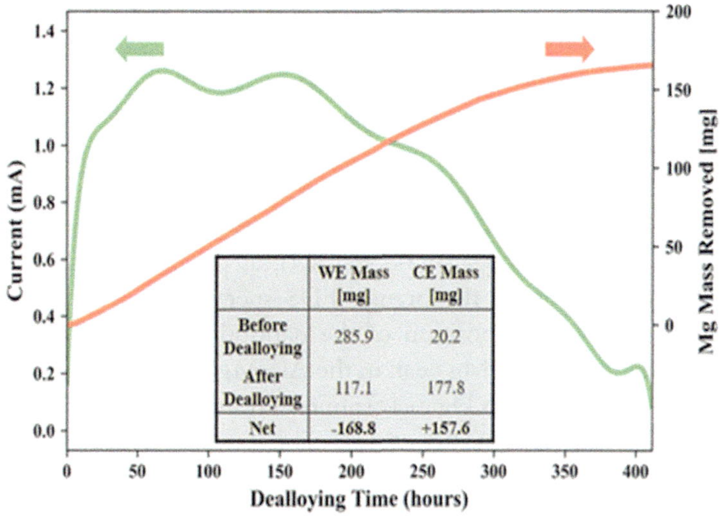

	WE Mass [mg]	CE Mass [mg]
Before Dealloying	285.9	20.2
After Dealloying	117.1	177.8
Net	-168.8	+157.6

Figure 6.3 Cell current profile associated with electrolytic dealloying at a constant potential of 450 mV versus Mg/Mg^{2+}(green, left); and corresponding mass of Mg removed during dealloying (orange, right). Summary of gravimetric electrode data (inset table).

Mg alloy was comparable to the mass of Mg calculated using the Faraday's laws of electrolysis [29].

The inset table in Fig. 6.3 contains these gravimetric data for the working and counter electrodes before and after dealloying. It can be seen that the decrease in mass of the starting Al—Mg working electrode after dealloying is comparable to the increase in mass of the Mg foil counter electrode after dealloying. These gravimetric data further show that Mg removed from the working electrode can be efficiently collected on the counter electrode. This Mg could be reused to fabricate a new Al—Mg parent alloy, which makes this novel dealloying process sustainable. Note that while a thick (~4 mm) working electrode was used in Fig. 6.3 to demonstrate how robust nanostructured bulk materials can be produced using this method, thinner electrodes can be used to significantly decrease dealloying time. Typically, samples with thicknesses in the range of a few micrometers can be dealloyed in about 45 hours, irrespective of the lateral dimensions of these samples. Indeed, it is important to realize that the lateral dimensions of the sample do not affect the dealloying time. Only the thickness is shown to have a critical impact on the dealloying time. In other words, for large-scale synthesis of nanoporous aluminum, at least one dimension of the sample should be in the range of a few micrometers. Therefore, in the subsequent Chapter 7, we propose the use of Al—Mg parent materials in the powder form to scale up this synthesis.

Dealloyed nanoporous Al was characterized by XRD, EDS, ICP, SEM, and TEM. The material was sealed in Kapton tape prior to XRD characterization in order to prevent it from reacting with air [4]. The orange pattern in Fig. 6.1b represents the typical XRD pattern of the nanoporous Al after dealloying. All the diffraction peaks in this pattern match with Al metal (PDF reference card number 01-085—1327), as depicted by the red stick pattern in Fig. 6.1b. Also, the orange EDS spectrum in Fig. 6.1c shows the typical elemental composition of the nanoporous Al after dealloying. It is seen that the dominant Mg peak in the Al—Mg parent alloy before dealloying (green EDS spectrum Fig. 6.1c) almost entirely vanishes after dealloying (orange EDS spectrum Fig. 6.1c). The content of residual Mg in nanoporous Al after dealloying was found to be 2 at. % from EDS data and 5.7 at. % from ICP data. Detailed EDS and ICP data are summarized in Table 6.1

Scanning and transmission electron micrographs of the dealloyed nanoporous Al are displayed at different magnifications in Figs. 6.4 and 6.5, respectively. Large macropores and macroligaments with average feature size in the range of ~0.5—1 μm are observed from SEM images at low magnifications as depicted in Fig. 6.4a and b. The large macropores come from the full dissolution of the (Mg) solid solution phase from the Al—Mg parent alloy. According to the Al—Mg equilibrium phase diagram, the maximum solubility of Al in the (Mg) phase near room temperatures is less than 2 at. %. Consequently, such a small amount of Al is etched away together with the (Mg) matrix, resulting in the large macropores and macroligaments. At intermediate (Fig. 6.4c) magnifications and high magnifications (Fig. 6.4d and e), it can be seen that the macroligaments are composed of ultrafine mesoscopic ligament/pore structures. These mesoscopic structures were characterized using TEM as seen in Fig. 6.5b and c; the average ligament size was found to be in the range 10—20 nm. The small nanoscale pores come from the selective dissolution of Mg from the β-$Al_{12}Mg_{17}$ phase.

6.5 Enhanced reactivity of monolithic nanoporous Al fuel with water as the oxidizer

In this section, we will study the reactivity of the dealloyed monolithic bulk NP-Al as a metal fuel, with neutral H_2O as the oxidizer, as we did

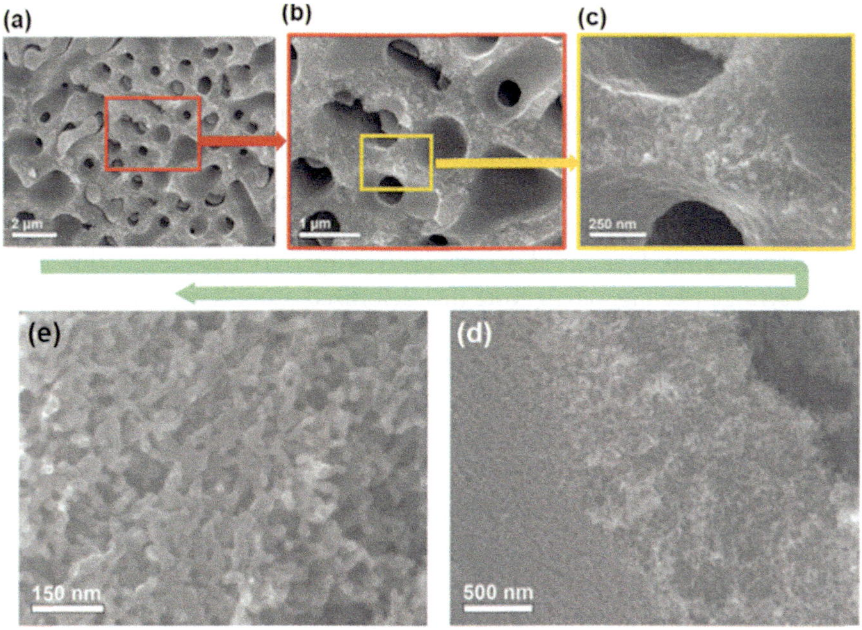

Figure 6.4 SEM of the fabricated hierarchical nanoporous Al at various magnifications. Large macropores and macroligaments with average feature size in the range of ~0.5–1 μm are observed at low magnifications (a,b). At intermediate magnifications (c) and at high magnifications (d,e), it is seen that the macroligaments are made of ultrafine mesoscopic ligament/pore structures.

Figure 6.5 TEM of nanoporous Al at various magnifications. (a,b) At low and intermediate magnifications the material is composed of a dense network of ligaments. (c) At high magnification, the characteristic ligaments size is in the range of 10–20 nm.

earlier with monolithic bulk NP-Zn in Chapter 4. The reaction of Al with H_2O produces hydrogen gas, Al hydroxide, and heat through Eq. (1.10), as discussed previously:

$$2Al\ (s) + 6H_2O\ (L) \rightarrow 2Al(OH)_3\ (s) + 3H_2\ (g) + heat \tag{6.1}$$

The amount and rate of hydrogen generated through the reaction of our hierarchical bulk nanoporous Al with water (Eq. 1.10) is quantitatively investigated using a customized Scion 456 gas chromatography (GC) system [4]. The schematic setup for hydrogen measurement is shown in Fig. 6.6a. In order to prevent rapid oxidation in air, the piece of bulk nanoporous Al used for hydrogen generation was brought in the reaction flask inside an argon-filled glovebox, and the flask was sealed before taking it outside the glovebox for GC measurements. Deionized water was then injected into the reaction flask using a syringe (see Fig. 6.6a). Note that deionized water is used to emphasize that our nanoporous Al-water system can produce hydrogen without any additives in the water. Thus, for practical application, any type of water could be used, including seawater. Helium was used as a carrier gas to transfer hydrogen produced from the reaction flask to the GC. The black, red, and blue curves in Fig. 6.6b show typical raw GC data associated with the amount of hydrogen produced 30, 185, and 380 minutes after injection of water, respectively. As time increases, the cumulative amount of hydrogen generated also increases. A similar trend was observed in Chapter 4 for the reaction of hierarchical nanoporous Zn in distilled water [4].

Fig. 6.6c shows the cumulative amount of hydrogen produced as a function of total reaction time. In the first 30 minutes, the hydrogen generation rate was found to be ~0.5 mg/minute per gram of nanoporous Al, which corresponds to ~5.3 mL/minute per gram of Al. The average hydrogen generation rate after 10 hours was about ~1 mL/minute per gram of Al. For the sake of comparison, the average hydrogen generation rate from hierarchical bulk nanoporous Zn in Chapter 4 is ~7.3 μg/minute per gram of Zn. This corresponds to 0.08 mL/minute per gram of Zn [4]. Therefore, it can be concluded that in the first 30 minutes, hydrogen generation rate by

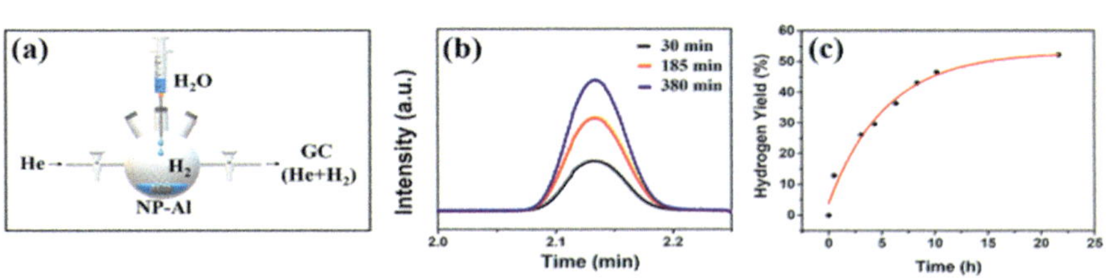

Figure 6.6 (a) Schematic of the setup for hydrogen measurement during GC experiments. (b) Typical raw GC data associated with the amount of hydrogen produced 30, 185, and 380 minutes after injection of water, respectively. (c) Cumulative amount of hydrogen produced after 22 hours.

hydrolysis of hierarchical nanoporous Al is more than 66 times faster than generating hydrogen by hydrolysis of hierarchical nanoporous Zn [4].

After a significant drop in the hydrogen generation rate of the Al-water system after ~22 hours, the hydrogen generation yield deduced from the total amount of hydrogen gas collected was found to be ~52% of the theoretical gravimetric capacity of this Al-water system (i.e., ~2 wt.% compared to 3.7 wt.%). For comparison, a yield of ~20% of the theoretical gravimetric capacity of the Zn-water system was achieved during hydrogen generation by hydrolysis of hierarchical nanoporous Zn in distilled water (i.e., 0.39 wt.% compared to ~2 wt.%) [4]. This remarkable performance (i.e., faster rate, higher capacity) of hierarchical nanoporous Al can be attributed to the nearly oxide-free nature of the surface of fabricated nanoporous Al (i.e., absence of Al_2O_3), as it will be clarified in the next section using XPS analysis. In contrast, the surface of nanoporous Zn was covered with residual oxygen species, because it was fabricated in air using an "aqueous" corroding medium [4].

Next, the reaction rate of our bulk nanoporous aluminum with water to produce hydrogen was further compared with the reaction rate of other reported Al-based materials systems in the literature. However, this is not a fair comparison because we do not include any additives in the aluminum-water system, unlike other reported works that use additives to enhance the reactivity of aluminum with water [36–38]. Producing hydrogen from aluminum and water without any additives presents several advantages for practical applications, as mentioned earlier. Data on hydrogen generation rate are shown in Table 6.2. Even though no additives are used in the present work, in general our reaction rate is competitive.

The hydrogen generation yield of ~52% measured above from GC data is an underestimated value because the hydrolysis process was stopped after 22 hours due to a significant drop in hydrogen generation rate. The sample was kept one more day in water to ensure the completion of the reaction. XPS characterization was then performed to investigate the oxidation state of Al in the solid $Al(OH)_3$ by-product formed through the reaction of hierarchical nanoporous Al with water. Note that due to the relatively small penetration depth in XPS, the sample was crushed into powder prior to XPS characterization in order to expose the inert parts of the material during XPS. The XPS survey spectrum reveals obvious O and Al peaks, as shown in Fig. 6.7a. In addition to O and Al, a small amount of Mg (amount: 7.2 at. % compared to 91.8 at. % of Al) was detected, suggesting the presence of a tiny fraction of residual Mg in the dealloyed nanoporous Al, in agreement with

Table 6.2 Literature data on hydrogen generation rate in Al-based systems with and without additives (nanoporous Zn is included because it was also discussed in this manuscript).

Material	Additive	Rate [g H_2/s/g material]	References
Bulk nanoporous aluminum	None	8.33×10^{-6}	This work
Ball milled and quenched activated aluminum	None	8×10^{-7}	[32]
Aluminum	γ-Al_2O_3	4×10^{-6}	[32]
Aluminum at 55°C	KCl, NaCl	2×10^{-4}	[32]
Aluminum	$Al(OH)_3$	3.4×10^{-4}	[33]
Aluminum	0.10 M Na_2SnO_3	No mass given	[34]
Aluminum	Ga–In, gallium (70:30, 7 wt%)	1.9×10^{-3}	[35]
Bulk nanoporous Zn	None	1.2×10^{-7}	[4]

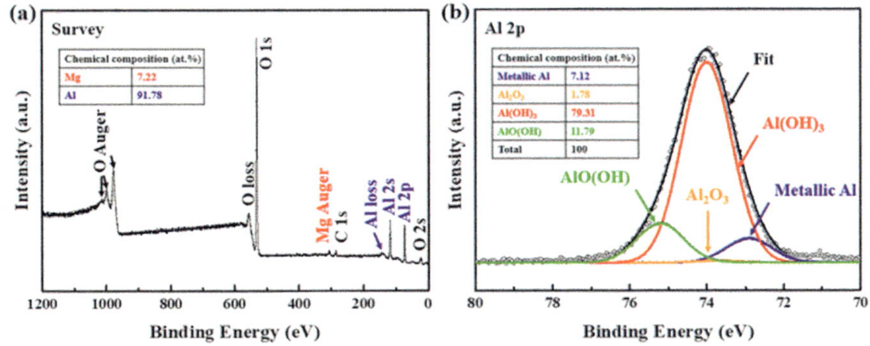

Figure 6.7 (a) XPS survey spectrum of the $Al(OH)_3$ solid by-product formed during the reaction of hierarchical nanoporous Al with water. (b) XPS high resolution core level spectrum of the Al 2p. The corresponding chemical composition is given in the inset table.

EDS and ICP data as discussed previously in Table 6.1. Fig. 6.7b displays a high-resolution Al 2p core level spectrum. As expected, the deconvolution of this Al 2p reveals the predominance of Al—OH peak at ~74 eV from the $Al(OH)_3$ compound [39]. In addition to $Al(OH)_3$, three other Al states were identified: metallic Al peak at ~72.9 eV, Al—O peak at ~73.7 eV from Al_2O_3, and finally, O=Al—OH peak at ~75.2 eV from the AlO(OH) compound [39,40]. The atomic fractions of all these compounds are shown in the inset table of Fig. 6.7b.

The fraction of Al in the form $Al(OH)_3$ was found to be 79.31 at. %, suggesting that the reaction yield during hydrolysis of Al to make hydrogen gas and $Al(OH)_3$ through Eq. (6.1) is at least 79.31%. It can be seen that 7.12 at. % of metallic Al is present, indicating that our hierarchical nanoporous Al has not fully reacted with water to produce hydrogen. The presence of unreacted aluminum can be rationalized by the 10—20 nm mesoscopic ligament size distribution in this hierarchical nanoporous Al. Based on our hypothesis stated earlier, the ideal size of Al for its full reaction with water is about ~7 nm. In the 10—20 nm ligament size distribution, ligaments near the lower limit (i.e., 10 nm) are susceptible to fully reacting with water, whereas ligaments near the upper limit (i.e., 20 nm) will only partially react with water. Therefore, the partially reacted ligaments account for the presence of metallic Al in the solid by-product. Further, an 11.79% fraction of AlO(OH) was detected and can be attributed to the gradual conversion of $Al(OH)_3$ into AlO(OH) when this material dehydrates during vacuum exposure and X-ray irradiation needed in XPS characterization. It should be emphasized that only traces of Al_2O_3 were found (1.78 at. %), suggesting that the nanoporous Al had not been exposed to oxygen during the fabrication process. The maximum hydrogen generation yield was deduced from the total fraction of $Al(OH)_3$ and AlO(OH) compounds, assuming that AlO(OH) evolves from $Al(OH)_3$. This gives a total yield of 91.10%.

6.6 Conclusions

The synthesis of hierarchical bulk nanoporous Al is illustrated through a unique nonaqueous electrolytic dealloying process that also allows for the simultaneous recovery of the sacrificial material (metallic Mg) selectively stripped from the parent alloy (Al–Mg alloy). The fabricated hierarchical nanoporous aluminum is made of macroscopic and mesoscopic ligament/pore structures with average characteristic size in the range of 10–20 nm for the mesoscopic ligaments. XPS data reveal that the nanoporous Al was not exposed to oxygen during its processing.

This material was used to produce hydrogen on-site with a yield ranging ~52%–90% by hydrolysis with deionized water, and without using any additives in the aluminum-water system to promote the reaction. The spontaneous combustion of bulk nanoporous aluminum in air and at ambient conditions was also found and indeed such materials are desirable as metal fuel. In terms of hydrogen generation by hydrolysis in water, a major advantage of the hierarchical porosity is that the large pores can facilitate transfer of water through the bulk of aluminum, allowing the mesoscopic ligaments to come in direct contact with water and react to produce hydrogen and heat. The large pores also facilitate the transfer of hydrogen gas out of the bulk of the material. This approach could be attractive for large-scale sustainable hydrogen production using only water and aluminum through the utilization of secondary aluminum or potentially carbon-free primary aluminum, which is currently being developed by Elysis (https://elysis.com/en).

Acknowledgments

Part of the content of this chapter has been published by J.S. Corsi, J. Fu, Z. Wang, T. Lee, A. K. Ng, and E. Detsi, ACS Sustainable Chemistry & Engineering 7, 71311194-11204 (2019), entitled, *Hierarchical Bulk Nanoporous Aluminum for On-Site Generation of Hydrogen by Hydrolysis in Pure Water and Combustion of Solid Fuels.* The authors are thankful to John Corsi, Jintao Fu, Zeyu Wang, Timothy Lee, Alexander Ng for their discussion and contributions.

References

[1] N. Phambu, Characterization of aluminum hydroxide thin film on metallic aluminum powder, Mater. Lett. 57 (2003) 2907–2913, https://doi.org/10.1016/S0167-577X(02)01395-2.

[2] E. Detsi, X. Petrissans, Y. Yan, J.B. Cook, Z. Deng, Y.-L. Liang, B. Dunn, S.H. Tolbert, Tuning ligament shape in dealloyed nanoporous tin and the impact of nanoscale morphology on its applications in Na-ion alloy battery anodes, Phys. Rev. Mater. 2 (2018) 055404, https://doi.org/10.1103/PhysRevMaterials.2.055404.

[3] H. Yaghoobnejad Asl, J. Fu, H. Kumar, S.S. Welborn, V.B. Shenoy, E. Detsi, In situ dealloying of bulk Mg_2Sn in Mg-ion half cell as an effective route to nanostructured Sn for high performance Mg-ion battery anodes, Chem. Mater. 30 (2018) 1815–1824, https://doi.org/10.1021/acs.chemmater.7b04124.

[4] J. Fu, Z. Deng, T. Lee, J.S. Corsi, Z. Wang, D. Zhang, E. Detsi, pH-controlled dealloying route to hierarchical bulk nanoporous Zn derived from metastable alloy for hydrogen generation by hydrolysis of Zn in neutral water, ACS Appl. Energy Mater. 1 (2018) 3198–3205, https://doi.org/10.1021/acsaem.8b00419.

[5] Z.M. Zhang, T. Lii, C.J. Xu, X.F. Guo, Microstructure of binary Mg-Al eutectic alloy wires produced by the Ohno continuous casting process, Acta Metallurgica Sinica (English Letters) 21 (2008) 275–281, https://doi.org/10.1016/S1006-7191(08)60049-6.

[6] E. Detsi, J.S. Corsi, Bulk Nanoporous Materials for On-Board Generation of Hydrogen and Other Products, U.S. Provisional Application, 2018, p. 972. No. 62/644.

[7] J. Erlebacher, M.J. Aziz, A. Karma, N. Dimitrov, K. Sieradzki, Evolution of nanoporosity in dealloying, Nature 410 (2001) 450–453, https://doi.org/10.1038/35068529.

[8] E. Detsi, Z. Vuković, S. Punzhin, P.M. Bronsveld, P.R. Onck, J.Th.M. De Hosson, Fine-tuning the feature size of nanoporous silver, CrystEngComm 14 (2012) 5402–5406, https://doi.org/10.1039/c2ce25313e.

[9] J. Weissmüller, K. Sieradzki, Dealloyed nanoporous materials with interface-controlled behavior, MRS Bull. 43 (2018) 14–19, https://doi.org/10.1557/mrs.2017.299.

[10] E.T. Lilleodden, P.W. Voorhees, On the topological, morphological, and microstructural characterization of nanoporous metals, MRS Bull. 43 (2018) 20–26, https://doi.org/10.1557/mrs.2017.303.

[11] Q. Chen, Y. Ding, M. Chen, Nanoporous metal by dealloying for electrochemical energy conversion and storage, MRS Bull. 43 (2018) 43–48, https://doi.org/10.1557/mrs.2017.300.

[12] M.B. Cortie, A.I. Maaroof, G.B. Smith, Electrochemical capacitance of mesoporous gold, Gold Bull. 38 (2005) 14–22, https://doi.org/10.1007/BF03215223.

[13] M.C. Tai, A. Gentle, M.D. Arnold, M.B. Cortie, Optical in situ study of de-alloying kinetics in nanoporous gold sponges, RSC Adv. 6 (2016) 85773–85778, https://doi.org/10.1039/C6RA18272K.

[14] Z. Zhang, Y. Wang, Z. Qi, W. Zhang, J. Qin, J. Frenzel, Generalized fabrication of nanoporous metals (Au, Pd, Pt, Ag, and Cu) through chemical dealloying, J. Phys. Chem. C 113 (2009) 12629–12636, https://doi.org/10.1021/jp811445a.

[15] J.R. Hayes, A.M. Hodge, J. Biener, A. V Hamza, K. Sieradzki, Monolithic nanoporous copper by dealloying Mn–Cu, J. Mater. Res. 21 (2006) 2611–2616, https://doi.org/10.1557/jmr.2006.0322.

[16] J.B. Cook, E. Detsi, Y. Liu, Y.-L. Liang, H.-S. Kim, X. Petrissans, B. Dunn, S.H. Tolbert, Nanoporous tin with a granular hierarchical ligament morphology as a highly stable Li-ion battery anode, ACS Appl. Mater. Interfaces 9 (2017) 293–303, https://doi.org/10.1021/acsami.6b09014.

[17] F. Meng, Y. Ding, Sub-micrometer-thick all-solid-state supercapacitors with high power and energy densities, Adv. Mater. 23 (2011) 4098–4102, https://doi.org/10.1002/adma.201101678.

[18] X. Lang, A. Hirata, T. Fujita, M. Chen, Nanoporous metal/oxide hybrid electrodes for electrochemical supercapacitors, Nat. Nanotechnol. 6 (2011) 232–236, https://doi.org/10.1038/nnano.2011.13.

[19] M. Hakamada, M. Mabuchi, Fabrication of nanoporous palladium by dealloying and its thermal coarsening, J. Alloys Compd. 479 (2009) 326–329, https://doi.org/10.1016/j.jallcom.2008.12.078.

[20] T.S. Dorofeeva, E. Seker, Electrically tunable pore morphology in nanoporous gold thin films, Nano Res. 8 (2015) 2188–2198, https://doi.org/10.1007/s12274-015-0726-x.

[21] E. Detsi, S. Punzhin, J. Rao, P.R. Onck, J.Th.M. De Hosson, Enhanced strain in functional nanoporous gold with a dual microscopic length scale structure, ACS Nano 6 (2012) 3734–3744, https://doi.org/10.1021/nn300179n.

[22] A. Wittstock, V. Zielasek, J. Biener, C.M. Friend, M. Bäumer, Nanoporous gold catalysts for selective gas-phase oxidative coupling of methanol at low temperature, Science 327 (2010) 319–322, https://doi.org/10.1126/science.1183591.

[23] S. Liu, J. Feng, X. Bian, J. Liu, H. Xu, The morphology-controlled synthesis of a nanoporous-antimony anode for high-performance sodium-ion batteries, Energy Environ. Sci. 9 (2016) 1229–1236, https://doi.org/10.1039/C5EE03699B.

[24] T.L. Maxwell, T.J. Balk, The fabrication and characterization of bimodal nanoporous Si with retained Mg through dealloying, Adv. Eng. Mater. (2017) 1–9, https://doi.org/10.1002/adem.201700519.

[25] S. Liu, J. Feng, X. Bian, Y. Qian, J. Liu, H. Xu, Nanoporous germanium as high-capacity lithium-ion battery anode, Nano Energy 13 (2015) 651–657, https://doi.org/10.1016/j.nanoen.2015.03.039.

[26] W. Ma, J. Wang, H. Gao, J. Niu, F. Luo, Z. Peng, Z. Zhang, A mesoporous antimony-based nanocomposite for advanced sodium ion batteries, Energy Storage Mater. 13 (2018) 247–256, https://doi.org/10.1016/J.ENSM.2018.01.016.

[27] S.H. Tolbert, E. Detsi, Hydrogen Storage in Nanoporous and Nanostructured Hydride Forming Metals, US Patent, 2013. US20180016139A1.

[28] W. Yang, X. Zheng, S. Wang, H. Jin, Nanoporous aluminum by galvanic replacement: dealloying and inward-growth plating, J. Electrochem. Soc. 165 (2018) 492–496, https://doi.org/10.1149/2.0881809jes.

[29] E. Detsi, M. Van De Schootbrugge, S. Punzhin, P.R. Onck, J.T.M. De Hosson, On tuning the morphology of nanoporous gold, Scripta Materialia 64 (2011) 319–322, https://doi.org/10.1016/j.scriptamat.2010.10.023.

[30] K. Ma, J.S. Corsi, J. Fu, E. Detsi, Origin of the volume contraction during nanoporous gold formation by dealloying for high-performance electrochemical applications, ACS Appl. Nano Mater. 1 (2018) 541–546, https://doi.org/10.1021/acsanm.7b00055.

[31] E. Detsi, M.S. Sellès, P.R. Onck, J.T.M. De Hosson, Nanoporous silver as electrochemical actuator, Scripta Materialia 69 (2013) 195–198, https://doi.org/10.1016/j.scriptamat.2013.04.003.

[32] J. Petrovic, G. Thomas, Reaction of Aluminum With Water to Produce Hydrogen, US Department of Energy, 2008, pp. 1–26.

[33] Y.K. Chen, H. Te Teng, T.Y. Lee, H.W. Wang, Rapid hydrogen generation from aluminum-water system by adjusting water ratio to various aluminum/aluminum hydroxide, Int. J. Energy Environ. Eng. 5 (2014) 1–6, https://doi.org/10.1007/s40095-014-0087-3.

[34] L. Soler, A.M. Candela, J. Macanás, M. Muñoz, J. Casado, Hydrogen generation from water and aluminum promoted by sodium stannate, Int. J. Hydr. Energy 35 (2010) 1038–1048, https://doi.org/10.1016/j.ijhydene.2009.11.065.
[35] A.V. Ilyukhina, A.S. Ilyukhin, E.I. Shkolnikov, Hydrogen generation from water by means of activated aluminum, Int. J. Hydr. Energy 37 (2012) 16382–16387, https://doi.org/10.1016/j.ijhydene.2012.02.175.
[36] H.Z. Wang, D.Y.C. Leung, M.K.H. Leung, M. Ni, A review on hydrogen production using aluminum and aluminum alloys, Renew. Sust. Energy Rev. 13 (2009) 845–853, https://doi.org/10.1016/j.rser.2008.02.009.
[37] B. Alinejad, K. Mahmoodi, A novel method for generating hydrogen by hydrolysis of highly activated aluminum nanoparticles in pure water, Int. J. Hydr. Energy 34 (2009) 7934–7938, https://doi.org/10.1016/j.ijhydene.2009.07.028.
[38] K. Mahmoodi, B. Alinejad, Enhancement of hydrogen generation rate in reaction of aluminum with water, Int. J. Hydr. Energy 35 (2010) 5227–5232, https://doi.org/10.1016/j.ijhydene.2010.03.016.
[39] Y.I. Seo, Y.J. Lee, D.-G. Kim, K.H. Lee, Y. Do Kim, Mechanism of aluminum hydroxide layer formation by surface modification of aluminum, Appl. Surf. Sci. 256 (2010) 4434–4437, https://doi.org/10.1016/j.apsusc.2010.01.011.
[40] R. Andreeva, E. Stoyanova, A. Tsanev, D. Stoychev, Influence of the surface pretreatment of aluminum on the processes of formation of cerium oxides protective films, in: Journal of Physics: Conference Series, 2016, https://doi.org/10.1088/1742-6596/700/1/012049.

CHAPTER SEVEN

Rapid synthesis of nanoporous aluminum powder by air-free electrolytic dealloying with recovery of sacrificial materials

7.1 Overview

So far, we have discussed the importance of producing nonprecious nanoporous metals such as nanoporous zinc (NP-Zn), nanoporous iron (NP-Fe), and nanoporous aluminum (NP-Al) on a large scale for use as metal fuels and mass production of hydrogen. In doing so, in Chapter 4, we presented the fabrication of NP-Zn by selective removal of Al from millimeter-sized arc-melted Zn-Al parent alloys through free corrosion dealloying, which usually takes a few days. In Chapter 5, we demonstrated that the free corrosion dealloying time to make NP-Zn can be reduced from a few days to just 4 minutes using as the parent alloy micrometer-sized Zn-Al powder particles produced by gas-atomization. This rapid synthesis enables the production of NP-Zn on a large scale. What about NP-Al? Can NP-Al be rapidly synthesized using the new air-free electrolytic dealloying concept we presented in Chapter 6. This concept was based on the selective electrolytic removal of Mg from a monolithic bulk Al-Mg parent alloy with a few millimeters thickness. Mg removal from such a thick sample took several weeks due to the diffusion-controlled nature of the process. Such a long dealloying time is not ideal for scaling up the process. Therefore, the present chapter aims at demonstrating the scalability of the air-free electrolytic dealloying process by significantly reducing the dealloying time of our original synthesis procedure described in Chapter 6, to obtain a twenty-fold increase in production capabilities. By doing so, first, we use an Al-Mg parent alloy in powder form instead of a monolithic bulk Al-Mg parent alloy for the rapid synthesis of NP-Al. In addition, since Mg is electrochemically etched away from the Al-Mg parent alloy using a Mg-ion conducting electrolyte, to further boost the dealloying rate, we improve the ionic conductivity of the Mg-ion conducting electrolyte by

Metals as Clean Fuels
ISBN: 978-0-443-13537-8
https://doi.org/10.1016/B978-0-443-13537-8.00002-6

introducing an additive and by investigating an alternative Mg-ion conducting electrolyte.

Next, we demonstrate the enhanced reactivity of the synthesized NP-Al powder for wet metal fuel oxidation by showing that the NP-Al powder can spontaneously react with water at near-neutral pH to produce hydrogen and aluminum hydroxide solid by-products with a much higher hydrogen yield (~70%–84%) than with bulk NP-Al (~52%).

Finally, for large-scale applications of NP-Al as a metal fuel, it is critical to demonstrate that spent NP-Al in the form of $Al(OH)_3$ can be easily recycled back to Al metal in industry. Knowing that Al smelting companies use Al_2O_3 as the feedstock to make Al metal by electrolysis, in this chapter we demonstrate that the $Al(OH)_3$ produced from the reaction between NP-Al and water can be readily converted to Al_2O_3 via simple heat treatments.

7.2 Rapid synthesis of nanoporous Al powder by air-free electrolytic dealloying with recovery of sacrificial Mg

In Chapter 6, we presented a new air-free electrochemical method to make NP-Al by selective electrolytic removal of Mg from a monolithic Al-Mg bulk parent alloy having a few millimeters thickness. Mg removal from such a thick alloy took several weeks since the process is diffusion-controlled. This long dealloying time is undesirable for the scalability of the process. Therefore, the present chapter is aimed at demonstrating the scalability of our process. In doing so, we modify the original synthesis procedure presented in Chapter 6 to obtain a twenty-fold increase in production capabilities. Specifically, the following three key modifications of the original synthesis procedure are performed for larger and faster batches:

(i) Instead of using a monolithic thick $Al_{30}Mg_{70}$ parent alloy as the working electrode, we used small-sized, crushed pieces (thus in the form of powder) of the same $Al_{30}Mg_{70}$ composition as in Chapter 6 in a mesh pouch to expose more surface area to the electrolyte during Mg removal. In a typical process, the $Al_{30}Mg_{70}$ parent alloy was created by melting pure Al (shots, Alfa Aesar, 99.9%) and pure Mg (chips, Sigma Aldrich, 99.98%) at 900°C in a graphite crucible using a tube furnace (OTF-1200X, MTI Corporation) with a quartz tube under argon gas flow. Next, the parent alloy (which is brittle in nature due to the presence of the $Al_{12}Mg_{17}$ intermetallic phase in Mg-rich Al-Mg alloys) was ground using a mortar and pestle in an argon-filled glove box into finer pieces near 1 mm in width. These pieces were loaded into a working electrode pouch made of stainless steel #120 mesh (0.125 mm openings).

(ii) The ionic conductivity of the APC electrolyte used in Chapter 6 was further improved by introducing 0.4 M of lithium chloride (LiCl powder, >99.98%, Sigma Aldrich) [1].

(iii) Instead of a single working and counter electrode, multiple alternating working and counter electrodes were used in parallel as shown in Fig. 7.1a and b. In the present work we demonstrate the use of two $Al_{30}Mg_{70}$ working electrodes (brownish arrows Fig. 7.1b) alternating with 3 Mg foils counter electrodes (blue arrows Fig. 7.1b). This configuration can be further scaled up for larger applications by increasing the number of alternating electrodes. Similar concepts in our design are

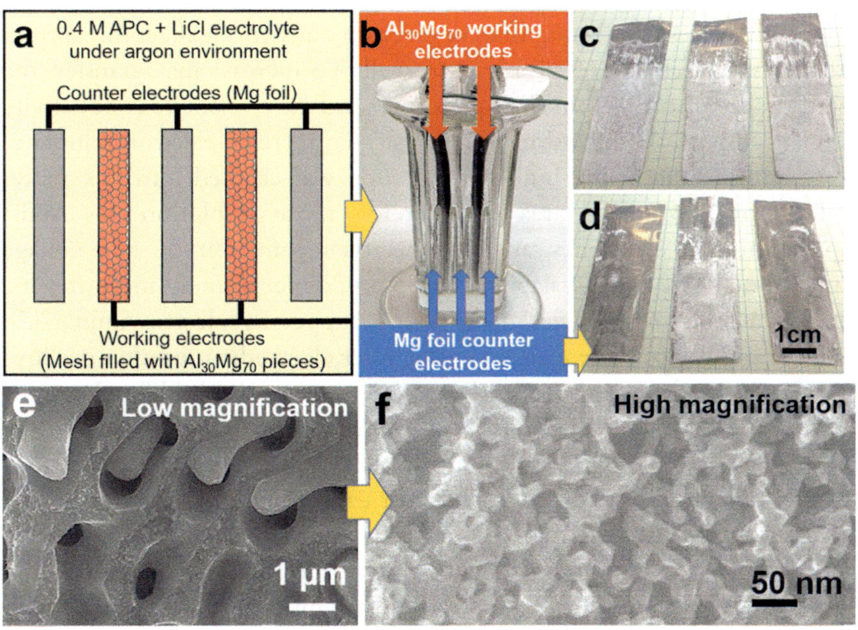

Figure 7.1 (a) Schematic of a large-scale dealloying chamber electrode layout. (b) Dealloying chamber before the addition of electrolyte. Crushed $Al_{30}Mg_{70}$ particles are loaded into a stainless steel mesh and used as the working electrode (brownish arrows); alternating Mg foils (blue arrows) serve as the counter electrodes. (c,d) Front and back sides of the 3 Mg foil counter electrodes after dealloying. Mg is plated on both sides of the middle electrode. (e) Low and (f) high magnifications SEM images of the hierarchical nanoporous aluminum morphology formed at the working electrodes.

used in commercial copper refining via electrowinning [2]. Similar to Chapter 6, the selective electrolytic removal of Mg was carried out using a Bio-Logic VMP-300 potentiostat inside an MBraun argon-filled glove box with H_2O and O_2 levels under 0.1 PPM. A constant potential of 0.45 V versus Mg/Mg^{2+} was applied. The Mg dissolution current is measured between the working and counter electrodes and monitored for completion in which NP-Al is collected and rinsed with THF for the next step.

Mg stripping from Al-Mg parent alloy and Mg plating onto the Mg foil counter electrode can be considered as a series of electrochemical steps: (1) dissolution of Mg ($Mg \rightarrow Mg^{2+} + 2e^-$) at the working electrode; (2) diffusion of Mg^{2+} ions in the electrolyte to the counter electrode, while electrons diffuse through the external circuit; (3) plating of Mg^{2+} ions at the counter electrode ($Mg^{2+} + 2e^- \rightarrow Mg$). This overall electrochemical reaction can be represented by the current measured between the working and counter electrodes arising from the diffusion of electrons through the external circuit. In general, processes affecting mass or electron transfer can affect this current and thus the electrode reaction rate as well. To increase mass transfer, the electrodes are placed within close proximity and the electrolyte conductivity is enhanced with the inclusion of LiCl salt. To increase electron transfer at the electrode surface, the bulk starting alloy was crushed into fine pieces to allow for more surface area and, therefore more dissolution sites. Additionally, multiple electrodes can be connected in parallel at the same voltage as mentioned earlier for scalability purpose. Copper refinery industries use similar concepts in their design to refine copper by electrowinning [2]. Fig. 7.1c and d shows the magnesium counter electrodes used after dealloying. Notice that the Mg is plated on both sides of the middle electrode, showing the beneficial additions of using multiple electrodes for increased counter-electrode surface area interactions. The peak current measured is larger in magnitude in comparison to our earlier publication [3], meaning that the reaction rate is occurring much faster with this configuration. Typical SEM images of the created hierarchical NP-Al are shown in Fig. 7.1e (low magnification) and 1f (high magnification). At low magnification, macropores and macroligaments with characteristic size in the micrometer range are observed (Fig. 7.1e). At high magnification, ligament/pore structures with characteristic size in the range of 10–20 nm are observed (Fig. 7.1f). These morphology and structure sizes are in agreement with our previous work [3], meaning that although the present dealloying current magnitude is much larger, the pore size of the produced hierarchical NP-Al was not changed. Using this new configuration, the rate of NP-Al production was ~5 mg per hour. Compared to our previous rate of NP-Al production of ~0.25 mg per hour, this improved configuration shows a twenty-fold increase in production capabilities [3]. Thus, production scalability for mass commercialization can be realized by following these three demonstrated principles.

7.3 Enhanced reactivity of nanoporous Al powder fuel with water as the oxidizer

In Chapter 6 (Section 6.5), we reported on the enhanced reactivity of the dealloyed monolithic bulk NP-Al as a wet metal fuel, using H_2O as the oxidizer. Similarly, this section aims to show that dealloyed NP-Al in powder form is even more reactive as a wet metal fuel than monolithic bulk NP-Al. In doing so, we use gas chromatography (GC) to quantitatively study the hydrogen production rate of NP-Al in water. NP-Al was sealed with rubber septum into a three-neck round bottom flask under an argon atmosphere to prevent oxidation. Deionized (DI) H_2O was then sparged with nitrogen gas for 10 minutes before injection via a syringe into the reaction flask. Helium carrier gas (99.9999% purity, Airgas) was flowed through the flask to push the gas product into GC. A Pulsed Discharge Detector (PDD) with Molecular Sieve 5A packed column was used to quantitatively determine the amount of hydrogen produced over time.

The cumulative amount of hydrogen produced as a function of the reaction time from NP-Al and H_2O is shown in Fig. 7.2. The first 10 data points (black square) were fit from experimental values over the course of 24 hours, and the rest of the curve (solid red line) was extrapolated for several days. The data profile matches the previous work presented in Chapter 6, emphasizing the reproducibility of these results. For 3 days of reaction time, the hydrogen generation yield from the NP-Al powder is ~70%, and for 7 days, the yield is ~84%. For the sake of comparison, the maximum yield extrapolated from the reaction of bulk NP-Al in Chapter 6 is only ~52%. In each case (i.e., NP-Al powder and bulk NP-Al), although the reaction is not predicted to have reached completion at this point, the reaction was visibly no longer producing hydrogen gas bubbles in solution.

This is due most likely to the formation of an $Al(OH)_3$ or Al_2O_3 envelope preventing further reaction of Al with DI H_2O [4–6]. It is important to note that our process for generating hydrogen gas from NP-Al is not optimized. We anticipate that a yield close to 100% can be achieved by further

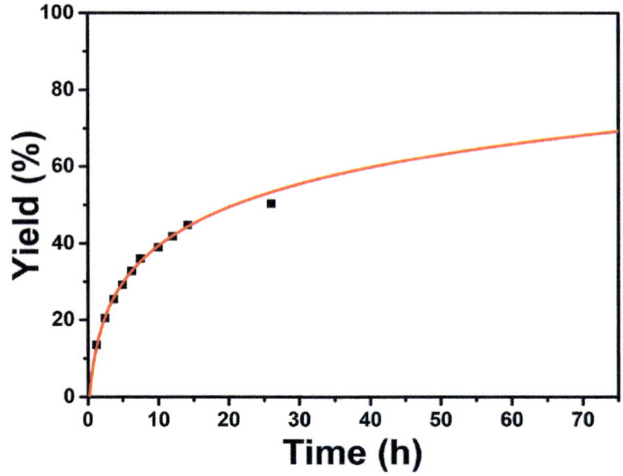

Figure 7.2 Black square: GC data associated with hydrogen yield from the reaction of hierarchical nanoporous aluminum with pure water. Red curve: Exponential fit extrapolated over 3 days.

reducing the size of the NP-Al powder. The latter can be achieved by using fine Al-Mg powder with a size in the submillimeter range as the parent alloy powder to make NP-Al). We also expect that varying the ratio of water to aluminum can cause variations in the reaction time length scale, as analyzed by Chen et al. [7] Surface area, temperature, pH, particle size effects, ultrasonic agitation, and adding other additives can lead to different hydrogen gas yield reaction rates as well [8—11].

7.4 Complete conversion of NP-Al into hydrogen gas, $Al(OH)_3$, and heat

The yields of ~70% and 84% obtained from GC measurements after 3 and 7 days, respectively, suggest that a fraction of the NP-Al has not reacted with water. Here, we demonstrate that complete conversion (100% yield) could be achieved by mechanically mixing/stirring the powder during the reaction. In doing so, the synthesized hierarchical NP-Al pieces were further ground (note that dealloyed nanoporous metals are brittle) from mm-sized pieces into a fine powder in the argon glove box using a mortar and pestle. The powder was brought into a round bottom flask together with a stir bar and sealed using a rubber septum before removal from the argon glove box. DI H_2O was flushed with argon gas for 10 minutes to remove dissolved oxygen from this water and then injected in excess into the round bottom flask with NP-Al. The ratio of water to aluminum used was 3 mL of H_2O per 100 mg of NP-Al for a large excess of H_2O. The injection of water resulted in the rapid formation of hydrogen gas, which increased the pressure in the sealed system. This pressure was relieved as necessary via a syringe. The flask was mechanically stirred for the desired length of time. Color change of the powder from black to white was observed after 3—4 days, indicating that most of the NP-Al has been converted into $Al(OH)_3$. Bubbles from hydrogen gas formation in the flask were no longer visibly observed after 6 days. After bubbling stopped, $Al(OH)_3$ was removed by centrifugation, rinsed with acetone, and then dried under vacuum overnight before further use in experiments. In the present work, the products studied are formed after reacting NP-Al with H_2O for 7 days; this represents a complete transformation as no more bubble formation is observed. An incomplete transformation was observed, where the products were also analyzed before the color change from black to white after only 3 days.

The XRD patterns of hierarchical NP-Al before and after its reaction with DI H_2O are presented in Fig. 7.3. The NP-Al is face-centered cubic (FCC), as shown by the black pattern. The corresponding $Al(OH)_3$ powder formed after the reaction is shown by the red pattern. This product exhibits a

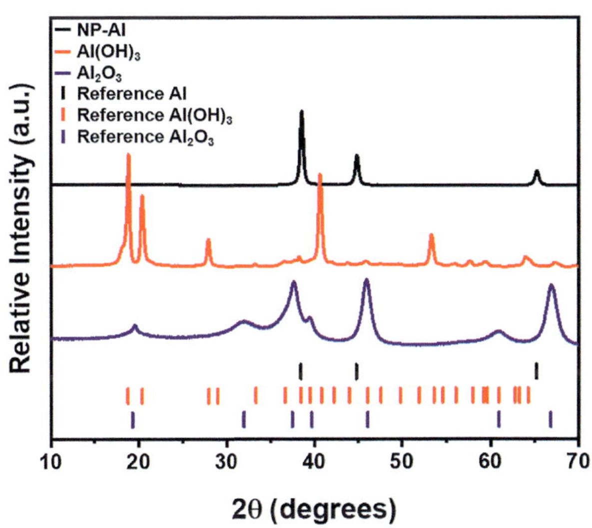

Figure 7.3 XRD patterns for hierarchical NP-Al (black), $Al(OH)_3$ (red), and Al_2O_3 (blue) with crystallography reference vertical stick patterns for FCC Al (black, PDF #00-004-0787), bayerite $Al(OH)_3$ (red, PDF #00-012-0457), and η-Al_2O_3 (blue, PDF #00-004-0875).

single phase of bayerite $Al(OH)_3$ distinguished by its first peak at 18.785 degrees as opposed to the gibbsite $Al(OH)_3$ phase, which has its first peak at 18.392 degrees [7]. The single-phase formation of bayerite observed agrees with other literature results of forming aluminum hydroxide from aluminum and water [12,13]. The formation of Al_2O_3 or other aluminum species, such as boehmite AlO(OH) was not observed by XRD from the reaction of NP-Al and DI H_2O.

Analysis of the incomplete product transformation of $Al(OH)_3$ shows two phases: the single bayerite $Al(OH)_3$ product and the original FCC Al. The data indicates an incomplete reaction, which is in agreement with GC results suggesting only a 70% reaction yield after 3 days. This data suggests that no other phases or Al species are detected during the transformation from Al to $Al(OH)_3$.

7.5 Conversion of $Al(OH)_3$ into activated alumina η-Al_2O_3 with high specific surface area

For large-scale application of Al as metal fuel, it is important to demonstrate that spent NP-Al in the form of $Al(OH)_3$ can easily be recycled back to Al metal in industry. Since Al smelting companies used Al_2O_3 as feedstock to make Al metal by electrolysis, in this section, we show that the $Al(OH)_3$ produced from the reaction between NP-Al and water can be readily converted to Al_2O_3 via simple heat treatments. Heat treatment triggers dehydroxylation of the aforementioned $Al(OH)_3$ powders to form

Figure 7.4 (a) Hierarchical NP-Al from large scale production and (b) corresponding activated Al_2O_3 powder obtained from $Al(OH)_3$ coproduced with H_2 by hydrolysis of the NP-Al from (a).

Al_2O_3 samples. The final product, a fine white powder, is shown in Fig. 7.4b in comparison to the starting black NP-Al pieces in Fig. 7.4a. The blue pattern in Fig. 7.3 shows the XRD results corresponding to the final Al_2O_3 product shown in Fig. 7.4b. The thermal decomposition of bayerite $Al(OH)_3$ to alumina has been known to transition through different metastable phases, forming η-Al_2O_3 at temperatures between 300 and 600°C, then θ-Al_2O_3 between 600 and 1100°C, and finally the thermodynamically stable phase of α-Al_2O_3 above 1100°C [13,14]. Thus, we see the single-phase formation of η-Al_2O_3 after heat treatments at 600°C. The diffraction peaks are reasonably broad regardless of reaction time because of the reduced crystallite size and lattice strain from the thermal decomposition process. Transformation to other phases outside of η-Al_2O_3 are possible using chemical additives; these transformation routes focus on lowering the activation energy for the nucleation of other transition Al_2O_3 phases [15,16]. These processes may be useful for future optimization procedures. Analysis of the XRD patterns for the incomplete transformation product of Al_2O_3 exhibits a mixture of alumina and unreacted aluminum. The lack of complete conversion of Al even during the open-air heating process to form Al_2O_3 is consistent with an inner composition of Al, protected by an outer layer of oxidized species [4–6].

EDS data showing the composition of the final Al_2O_3 product are summarized in Table 7.1. An Al-to-O ratio of ~0.689 is deduced from these EDS data, which is very close to the Al-to-O ratio of 0.667 expected for Al_2O_3. A ~2 at. % residual Mg remains in the sample due to incomplete dealloying. XPS characterization data are shown in Fig. 7.5. The deconvoluted Al 2p peak in Fig. 7.5a reveals that ~95.23% of Al is in the form of Al^{3+} from Al_2O_3, detected at 74.5 eV, and some ~4.77% metallic Al detected at 72.7 eV [17,18]. It is expected that some unreacted Al will be detected because the extrapolated reaction yield is only 84% by GC; however, the presence of metallic Al near the surface might be minimized compared to the bulk. The deconvoluted O 1s peak in Fig. 7.5b suggests that ~94.66% of the near-surface oxygen atoms are involved in O-Al bonds from Al_2O_3, with the peak at 531.3 eV, while ~5.14% of the near-surface

Table 7.1 Typical EDS data showing the chemical composition of activated Al_2O_3 obtained by heat treatment of the $Al(OH)_3$ hydrolysis byproduct.

Element	Al	Mg	O
Composition (at. %)	40	2	58

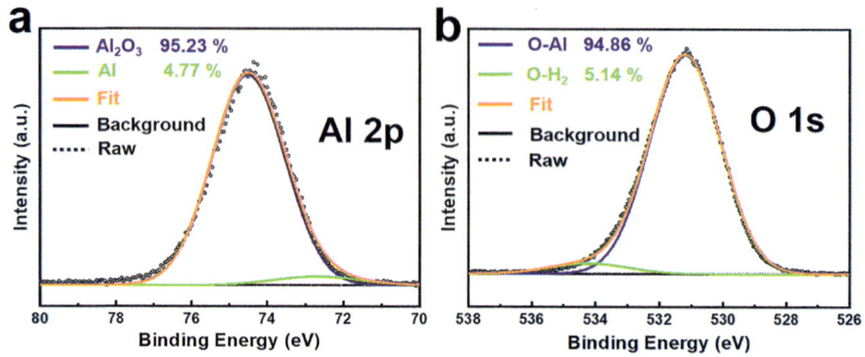

Figure 7.5 High resolution XPS spectra of Al 2p (a) and O 1s (b) peaks obtained from activated alumina. In (a) ~95.23% of aluminum has been converted into Al_2O_3, with only ~4.77% of unreacted residual Al. In (b) ~5.14% of near-surface oxygen atoms come from water molecules postadsorbed in the pores of activated alumina.

oxygen atoms come from adsorbed water molecules, with peak at 534.7 eV [19,20]. No peak associated with $Al(OH)_3$ was detected at ~532.4 eV, indicating the full conversion from $Al(OH)_3$ to Al_2O_3 [19]. The presence of adsorbed water molecules near the surface of activated alumina as suggested from XPS data is interesting; indeed, since the samples were activated at 600°C for 12 hours, all near-surface water molecules and hydroxyl groups were removed during this heat treatment step. Consequently, water molecules near to surface of activated Al_2O_3 can only come from postadsorption of moisture, when these samples are exposed to air. This good adsorption capability of activated alumina justifies its usage in applications such as water filtration for fluoride treatment, hydrocarbon impurity capture in the automobile and petroleum industries, or as a desiccant for dry gases [21–23].

The typical nanoscale morphology of the activated η-Al_2O_3 product as obtained from TEM characterization is shown at different magnifications in Fig. 7.6. The material exhibits an ultrafine nanoporous morphology, with ligaments in the range of ~2–5 nm and pores in the range of ~4–6 nm, which are much smaller than the ~10–20 nm ligament/pore structures present in the starting NP-Al before the hydrolysis reaction to produce H_2 and $Al(OH)_3$. As aforementioned, the origin of these pores are from the removal of hydroxide groups from the $Al(OH)_3$ precursor in the form of H_2O, and not from the nanostructuring created during electrochemical dealloying, which justify why the characteristic pore size in activated alumina (4–6 nm) is much smaller than in NP-Al (10–20 nm). A line scan EDS across a bright field STEM image (not shown) of the activated

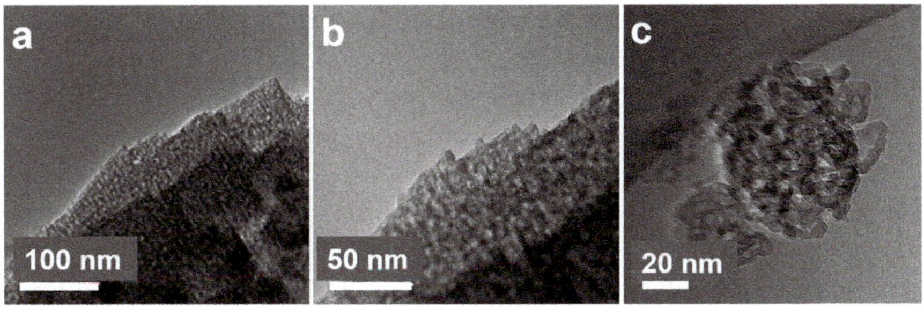

Figure 7.6 TEM images of activated Al_2O_3 at different magnifications. A nanoporous morphology with a characteristic pore size of about ~4–6 nm is observed at high magnification.

η-Al_2O_3 product reveals that the elements detected are Al, O, and residual Mg in agreement with the EDS results from Table 7.1.

To further confirm the nanoporous morphology observed in TEM images (see Fig. 7.6), nitrogen porosimetry measurements were performed on the activated η-Al_2O_3 product. The adsorption (black) and desorption (red) isotherms are shown in Fig. 7.7. The isotherm is a combination of a Type IV isotherm (up to $P/P_0 = 0.8$) typical of mesopores and a Type III isotherm (from $P/P_0 = 0.8-1.0$) typical of macropores [24]. Using these curves, the BET method was used to obtain the surface area of the activated Al_2O_3 samples, which was determined to be 161.2 m^2/g. The BJH adsorption average pore width is 6.3 nm, while the BJH adsorption volume-weighted pore size distribution shows a peak at 6.1 nm. The hysteresis in the adsorption and desorption isotherms from 0.4 to 0.8 P/P_0 (Type H1/H2) indicates that the pores have a smaller neck size around ~6.2 nm

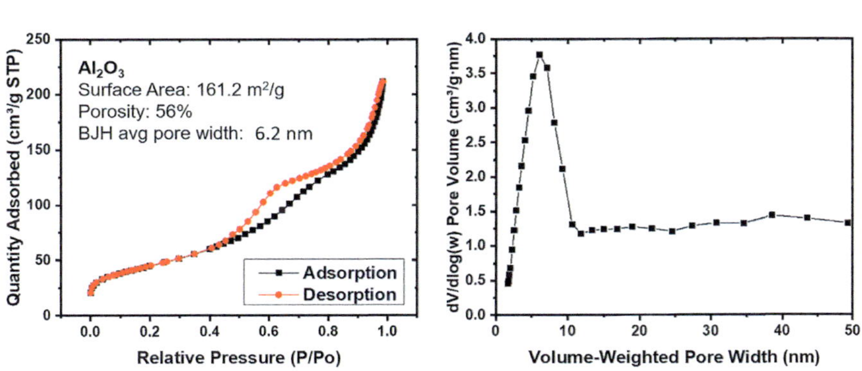

Figure 7.7 Adsorption and desorption isotherms (left) and BJH adsorption volume-weighted pore size distribution (right) of activated Al_2O_3.

(BJH desorption), while the hysteresis above 0.8 P/P_0 (Type H3) could be due to the ligament structure [24].

These results are in agreement with the characteristic pore size obtained from TEM. Nitrogen porosimetry measurements performed on the incompletely converted Al_2O_3 sample (data not shown) indicate that the partially activated Al_2O_3 exhibits lower porosity, stemming from the lower reaction yield in producing $Al(OH)_3$ from hierarchical NP-Al. While the 161.2 m^2/g measured in the fully converted η-Al_2O_3 is high, it should be mentioned that this value is relatively lower than the specific surface area of commercial activated γ-Al_2O_3 from Alfa Aesar (Catalog no. 43,832, catalyst support, high surface area, bimodal), as shown in Table 7.2. Thus, optimization processes to maximize the specific surface area like those used presumably in industry for commercial activated alumina products are necessary for further improvements to the hydrogen gas production value proposition.

7.6 Search for new electrolytes for the fabrication of ultrafine nanoporous Al

Although the air-free electrolytic dealloying method presented in this chapter and the previous chapter is successfully used to create NP-Al with ligament size in the range of ≈10–20 nm by selective electrolytic removal of Mg from Al-Mg parent alloys, [3] it should be acknowledged that this novel dealloying method faces drawbacks in flexibility because the number of available Mg-ion conducting electrolytes in the literature is limited.

Indeed, the most common nonaqueous electrolyte of choice is 0.4 M all-phenyl complex or "APC" used by the Mg-ion battery community [25–27], which is derived from in-situ reaction between two equivalents of Lewis base phenylmagnesium chloride (C_6H_5ClMg) and one equivalent of Lewis acid aluminum trichloride ($AlCl_3$) in tetrahydrofuran (THF) solvent. To our knowledge, APC is also the only electrolyte tested so far for air-free electrolytic dealloying with Mg as the sacrificial component [3].

Table 7.2 Comparison of BET surface area for different activated Al_2O_3 samples.

	Incomplete Al_2O_3	**Activated Al_2O_3**	**Commercial (Alfa)**
BET surface area (m^2/g)	95.47	161.2	220–260
BJH adsorption average pore width (nm)	7.1	6.2	9
Porosity	39.8%	56.4%	–

Since this chapter aims at demonstrating the scalability of the air-free electrolytic dealloying process to make NP-Al, we also need to demonstrate that new electrolytes other than APC can be used to make NP-Al. Therefore, herein we show how NP-Al can be created via air-free electrolytic dealloying by using a glyme-based Mg-ion conducting electrolyte made of 0.25 M $MgCl_2/AlCl_3/Mg(TFSI)_2$ (1:2:1 M ratio) in 1,2-dimethoxyethane (DME) solvent called "MACT."

This electrolyte was recently developed by Yang et al. for use in Mg-sulfur batteries and chosen for its relatively high ionic conductivity (6.82 mS/cm) [28]. We also show how this electrolyte can be used to create NP-Al with much finer ligaments and pores compared to APC. Analysis of NP-Al made in both APC as the control and MACT electrolytes shows that the main advantages of electrolytic dealloying mentioned earlier are still present, including the sustainable recovery of the sacrificial component, and the high reactivity from the nearly oxide-free NP-

Al created. This high reactivity was tested by reacting NP-Al created in MACT electrolyte with deionized (DI) H_2O, which undergoes a hydrolysis reaction that can produce hydrogen and various high-temperature alumina byproducts and may have applications in power generation or solid rocket fuel additives.

The performance of both APC and MACT as Mg-ion electrolytes was first investigated to show that both electrolytes can be used for electrolytic dealloying. In doing so, these electrolytes were characterized using electrochemical impedance spectroscopy (EIS) and cyclic voltammetry (CV), as shown in Fig. 7.8. EIS is first conducted at OCV for the MACT and APC electrolytes using symmetrical Mg electrodes with geometrical area

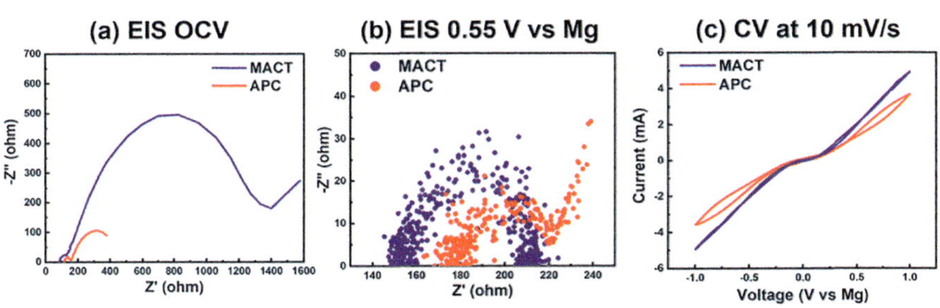

Figure 7.8 (a,b) EIS comparing MACT and APC electrolytes at (a) OCV, and (b) 0.55 V versus Mg using scatter plot overlay of four consecutive scans; (c) typical cyclic voltammetry tests from −1 to 1 V versus Mg at 10 mV/second using symmetrical 1 cm^2 Mg electrodes.

of 1 cm^2 (Fig. 7.8a), The electrolyte resistance (R_Ω) was then estimated from the real axis Z' value to be lower for MACT ($R_\Omega \approx 84\ \Omega$) compared to APC ($R^{\sim}116\ \Omega$). Next, the charge-transfer resistance (R_{CT}) at OCV estimated from the semicircle diameter was found to be much larger for MACT ($R_{CT} \approx 1300\ \Omega$) compared to APC (RCT $\approx 300\ \Omega$). Since electrolytic dealloying will occur under an applied potential of 0.55 V versus Mg, EIS was also conducted at the dealloying potential of 0.55 V versus Mg as shown in Fig. 7.8b, where four consecutive scans are combined in a scatter plot.

Here MACT still has a lower R_Ω value of $\approx 154\ \Omega$ compared to $\approx 178\ \Omega$ for APC, but the values of R_{CT} are more comparable for APC and MACT. The impact of these resistive properties is seen in the CV performance of the cell.

Fig. 7.8c shows the typical CV collected between −1 and 1 V versus Mg at the scan rate of 10 mV/second using symmetrical Mg electrodes with the geometrical area of 1 cm^2 Mg in MACT (blue) and APC (red). Note that the current measured, which arises from Mg stripping (oxidation) and plating (reduction), is greater in MACT compared to APC for the symmetrical Mg cell at 0.2 V versus Mg and higher. Thus, at the relevant dealloying potential of 0.55 V versus Mg, which has a sufficient overpotential to drive the reaction, MACT will be useable in electrolytic dealloying and may also have improved electrochemical performance due to its lower R_Ω value.

Next, we use both the APC and MACT electrolytes to dealloy Al-Mg parent alloys and create NP-Al. A parent alloy with composition $Al_{30}Mg_{70}$ at. % was created by melting the underlying Al and Mg at 850°C using a tube furnace under argon flow. Before dealloying, the $Al_{30}Mg_{70}$ at. % parent alloy, which is very brittle, was crushed into coarse powder and placed inside a stainless steel mesh pouch to use as the working electrode. The decision to use a powder increases the dealloying kinetics and reaction rate by increasing the surface area of the working electrode.

This $Al_{30}Mg_{70}$ at. % composition is chosen because it contains two phases: hexagonal close-packed Mg and $Al_{12}Mg_{17}$ phases as shown by the black X-ray diffraction (XRD) curve in Fig. 7.9a [3]. The reference and counter electrodes used are Mg foils, and the electrolyte used is APC or MACT to create two different types of NP-Al samples. During electrolytic dealloying, a voltage of 0.55 V versus Mg is applied between the working and reference electrodes, which causes Mg to be removed from the working electrode and plated onto the Mg foil counter electrode. This stripping of

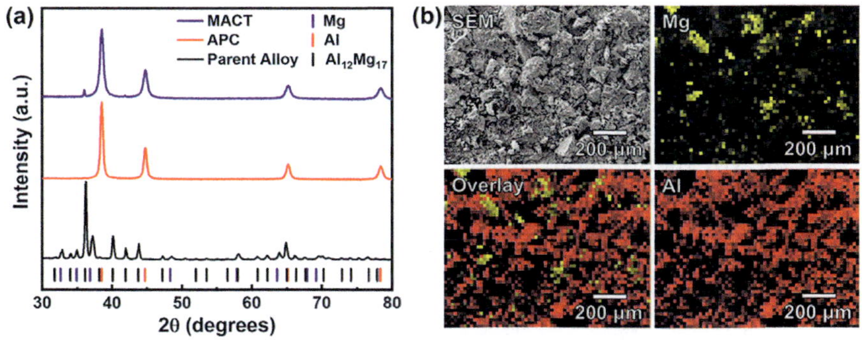

Figure 7.9 (a) XRD characterization before and after dealloying for NP-Al samples created in APC and MACT electrolyte; (b) EDX mapping characterization of NP-Al sample created in MACT electrolyte after dealloying. (Al PDF # 01-085-1327, $Al_{12}Mg_{17}$ PDF # 01-073-1148, Mg PDF # 00-004-0770).

Mg leaves behind NP-Al at the working electrode. The corresponding XRD of NP-Al is shown in Fig. 7.9a by the blue and red curves for MACT and APC respectively. After removing Mg from the parent alloy, both samples have face-centered cubic Al as the remaining dominant phase. However, the MACT sample still has a small peak around $2\theta\tilde{}36$ degrees indicating residual $Al_{12}Mg_{17}$. Energy dispersive X-ray spectroscopy (EDX) was also used to characterize the elemental composition of the NP-Al samples. A composition of 97 at. % Al and 3 at. % Mg was detected for the APC sample, while a composition of 85 at. % Al and 15 at. % Mg was detected for the MACT sample.

Further characterization of the elemental composition for the MACT sample via EDX mapping is shown in Fig. 7.9b with Al in red and Mg in yellow. Note that this sample is not fully homogeneous as regards the distribution of Al and Mg concentrations after dealloying; most regions in EDX mapping are Al-rich in red (EDX: 94 at. % Al, 6 at. % Mg), while some residual Mg-rich regions in yellow are detected (EDX: 44 at. % Al, 56 at. % Mg). These two types of regions (i.e., Al-rich, and Mg-rich regions) balance to bring the EDX of the total region to 85 at. % Al and 15 at. % Mg as mentioned above. The residual Mg-rich regions observed after dealloying can be justified by the presence of partially dealloyed powder particles, caused by the relatively large size of some powder particles (by the time small particles are nearly fully dealloyed, large particles are not), in combination with the dense nature of the NP-Al created in MACT electrolyte (as will be discussed in the next paragraph), which makes it difficult for dissolved

Mg to diffuse out of the bulk of parent materials with large particle size. Since these Mg-rich regions are highly localized, the residual Mg does not impact the overall performance of the NP-Al material.

To characterize the structure size of the NP-Al, scanning electron microscopy (SEM), transmission electron microscopy (TEM), and small-angle X-ray scattering (SAXS) are used as shown in Fig. 7.10. Here SEM shows

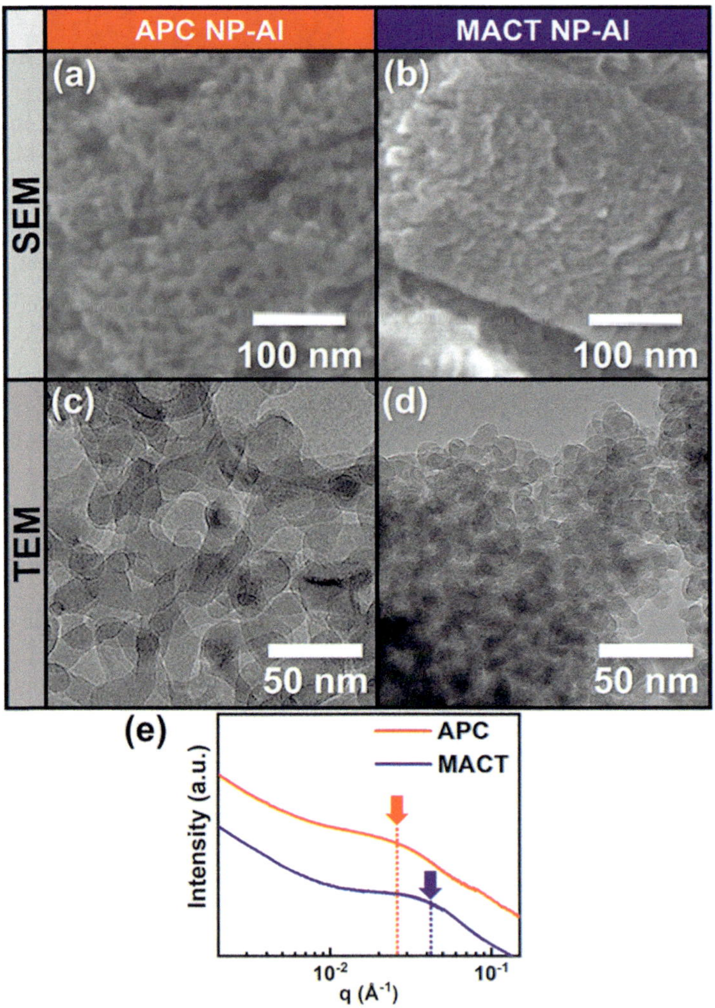

Figure 7.10 Nanostructural characterization showing (a,b) SEM of NP-Al samples created in APC and MACT respectively, and (c,d) TEM of NP-Al created in APC and MACT respectively. (e) SAXS data with estimated peak q_{max} values identified by dotted lines and arrows.

the nanoporous morphology of the NP-Al samples made in the APC and MACT electrolytes in Fig. 7.10a and b. The pores of NP-Al made in MACT are visibly smaller than those for NP-Al made in APC Higher magnification to examine the ligaments using TEM is shown in Fig. 7.10c and d. For NP-Al made in APC, the typical dealloyed nanoporous structure with bicontinuous and well-defined ligaments and pores is observed; this is similar to other types of dealloyed systems like nanoporous gold [29,30] and matches previous NP-Al structures dealloyed in APC as well [3]. However, the NP-Al ligaments and pores appear to be smaller and more dense in the MACT sample. To estimate the ligament size, ImageJ software was used on TEM images for both APC and MACT samples as shown in Fig. 7.11.

Here, the ligament size for the APC sample ranges between 10 and 20 nm in agreement with previous results [3], while the ligament size for MACT samples is smaller at ≈6−14 nm. To analyze the ligament structure size in a more quantitative way, SAXS is used as shown in Fig. 7.10e. Note that SAXS measurements give a more comprehensive overview of the morphology of the entire sample compared to microscopy since the probed area in SAXS here is 2 mm^2, which is larger than the typical 0.04 μm^2 area viewed in TEM. Here we see in SAXS that the peak denoting the averaged distance between ligaments shifts to higher q values when comparing NP-Al

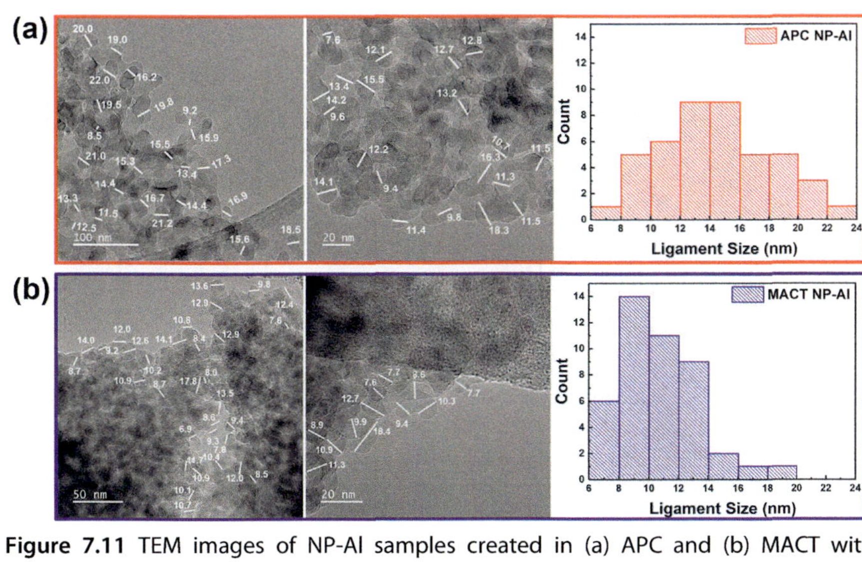

Figure 7.11 TEM images of NP-Al samples created in (a) APC and (b) MACT with selected ligaments, measured using ImageJ software and plotted in the histogram.

created in APC versus MACT. This means that the averaged distance between NP-Al ligaments d in the MACT sample is smaller than that of the APC sample given the approximate relationship for dealloyed nanoporous structures used in Eq. (7.1) [30,31]:

$$q_{max} \approx 1.23 \frac{2\pi}{d} \quad (7\ 1)$$

Here q_{max} is estimated for NP-Al created in APC and MACT as ≈ 0.026, and ≈ 0.042 Å^{-1} denoted by the red and blue vertical dotted lines respectively, which corresponds to an averaged distance d between ligaments of ≈ 30 nm for APC (corresponding to ≈ 15 nm ligament sizes) and ≈ 18 nm for MACT (≈ 9 nm ligament sizes) [30,31]. Thus, much smaller ligament and pore sizes can be achieved by switching from APC to MACT electrolytes for electrolytic dealloying.

After compositional and morphological characterization of both NP-Al samples, we reexamine some of the main advantages of electrolytic dealloying for the MACT sample. First, the recovery of the sacrificial component in dealloying is important for sustainable and practical applications [32]. Here we show recovery of Mg at the counter electrode in Fig. 7.12a and b using both APC and MACT electrolytes. Interestingly, APC recovered Mg exhibits a solid, single formation, while MACT recovered Mg is in the form of a dark gray, loose powder. XRD analysis in Fig. 7.12c of the two types of recovered Mg at the counter electrode (CE) show a pure Mg phase for

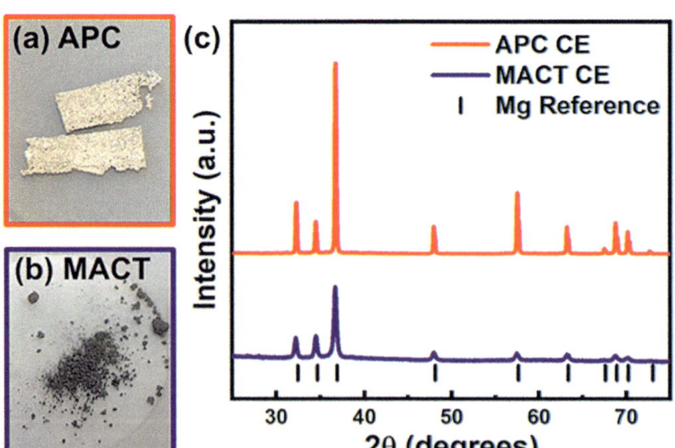

Figure 7.12 Mg recovery process after dealloying showing (a) solid Mg from APC, (b) powder Mg from MACT, and (c) XRD characterization of both recovered Mg forms.

samples created in both electrolytes, albeit a more crystalline result for APC is noticed compared to MACT. Furthermore, the microstructure of the MACT recovered Mg powder was analyzed and shown in Fig. 7.13. Here, ligament sizes can be seen on the order of 50–200 nm. Because of the nanoscale feature size of the Mg powder recollected at the counter electrode, there may be other possible applications for this value-added byproduct that can potentially increase the productivity of electrolytic dealloying in MACT via synthesis of two desired products.

Finally, the high reactivity of NP-Al made in MACT is investigated by reacting it with DI H_2O as shown in Fig. 7.14. The reaction chamber used is kept under argon, and the water is sparged with argon for 15 minutes prior to the reaction to remove oxygen as an oxidizer. Immediately upon reaction a large metal fire occurs between NP-Al made in MACT and H_2O as shown in Fig. 7.14a and c. After the first 5 minutes of the reaction, the entire solid sample is collected via filtration as shown in Fig. 7.14d. Here several solid white reaction byproducts are observed (blue circles). When analyzing these white byproducts by XRD in Fig. 7.14e (blue curve), we detect the Al phase from unreacted NP-Al and various α-, γ-, or η-phases of Al_2O_3. Meanwhile, when NP-Al made in APC reacts with DI H_2O for 3 days, no large metal fire is observed in the initial reaction, where hydrogen is released and collected relatively slowly and bayerite $Al(OH)_3$ is detected (red curve) [19]. In the reaction between Al and H_2O, three types of reactions are possible based on the experimental conditions:

$$2Al + 6H_2O \ (< 280^{\circ}C) \rightarrow \ 2Al(OH)_3 + 3H_2 + heat \tag{7.2}$$

Figure 7.13 SEM of MACT counter electrode at (a) low-magnification and (b) high-magnification.

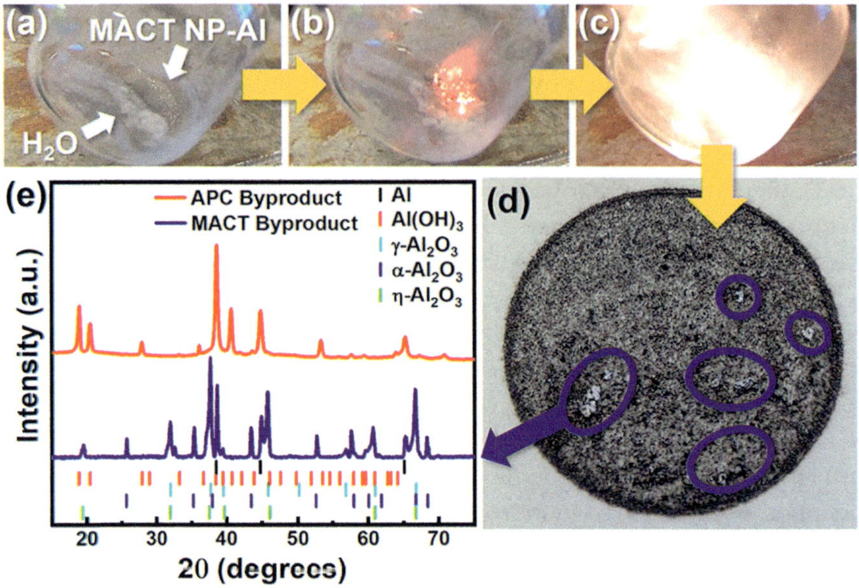

Figure 7.14 (a–c) Reaction between NP-Al created in MACT and DI H_2O showing the development of initial fire; (d) collection of all solid byproducts after 5 minutes with some white byproducts circled in blue; (e) XRD of byproducts from the reaction of NP-Al created in APC (red curve, after 3 days) and MACT (blue curve, after 5 minutes) with water. $Al(OH)_3$ PDF # 00-012-0457, γ-Al_2O_3 PDF # 00-002-1420, α-Al_2O_3.

$$2Al + 4H_2O \ (280-480°C) \rightarrow \ 2AlO(OH) + 3H_2 + heat \tag{7.3}$$

$$2Al + 3H_2O \ (> 480°C) \rightarrow \ Al_2O_3 + 3H_2 + heat \tag{7.4}$$

Note that heat is released because all three of these reactions are highly exothermic and that some or all reactions may occur to varying degrees due to varying high temperatures reached from exothermic self-heating [4,34]. To produce the Al_2O_3 phases detected in the NP-Al byproduct, the Al-containing byproducts bayerite $Al(OH)_3$ and boehmite $AlO(OH)$ from Eq.(7.2) and Eq. (7.3) can convert to η- and γ-phases of Al_2O_3 respectively from $\approx$300–600°C [35,36], while α-Al_2O_3 is formed at very high temperatures above 1100°C (although some reports have shown nanocrystalline α-Al_2O_3 can form at as low as 300–400°C under certain circumstances) [34,36]. We hypothesize that this wet oxidation reaction creating Al_2O_3, which becomes self-supported from the large exothermic reaction noticed, only occurs for the NP-Al created in MACT but not for the NP-Al created in APC due to the more dense nanoporous structure observed with smaller ligament sizes of the NP-Al created in MACT. Further optimization of reaction parameters such as the amount of NP-Al powder, amount of water, or even the size and shape of container used can further improve the exothermic self-heating and the yield for Al_2O_3 formation via Eq. (7.4).

Such optimization would be preferable to increase the wet capacity of hydrogen in respect to aluminum and water, which is 3.73 wt.% for Eq. (7.2) and 5.60 wt.% for Eq. (7.4). Furthermore, a faster reaction may occur for Eq. (7.4) as the full yield for Eq. (7.2) can take several days to occur [33]. Thus, further studies optimizing the reaction of NP-Al created in MACT with water or other strong oxidizers can unlock potential applications for rapid oxidation of Al and release of hydrogen for high-power fuel purposes like for space propulsion [37].

In summary of Section 7.6, NP-Al can be synthesized by using an alternative electrolyte besides APC, such as MACT, which improves the versatility of the electrolytic dealloying method. At the applied relevant dealloying potential of 0.55 V versus Mg, MACT exhibits better electrochemical performance than APC and can be used to synthesize a denser nanoporous structure with ligament sizes ≈6–14 nm. Furthermore, the advantages of electrolytic dealloying remain present. Sacrificial Mg used during dealloying in MACT electrolyte to create NP-Al can also be recovered, thus making this synthesis method more sustainable and practical for scaling purposes. Finally, the high reactivity of dense NP-Al created in MACT exhibited by its exothermic reaction with DI H_2O produces a large amount of initial heat that can form various phases of Al_2O_3, thus warranting further exploration of the application of NP-Al created in MACT for high-power fuel purposes.

7.7 Conclusions

Important issues as regard applications are scalability and value-added byproduct of hydrogen gas production via the hydrolysis reaction of nanoporous aluminum with pure water. Chapter 7 demonstrates that a 20-fold improvement in the dealloying rate to produce nanoporous aluminum was found using various strategies, ranging from increasing the ionic conductivity of the electrolyte to the number of electrodes simultaneously used in the cell.

It is relevant to note that sacrificial Mg used during dealloying in MACT electrolyte to create NP-Al can also be recovered, thus making this synthesis method more sustainable and practical for scaling purposes.

Finally, the high reactivity of finer NP-Al created in MACT exhibited by its exothermic reaction with DI H2O produces a large amount of heat that can form various phases of Al_2O_3, thus warranting further exploration on the application of NP-Al created in MACT for high-power fuel purposes. In particular, for commercial viability of the activated η-Al_2O_3 product produced via hydrolysis of nanoporous Al systems, further optimization toward even higher surface areas is recommended for future applications of this reaction process.

Acknowledgments

Part of the content of this chapter has been published by Timothy Lee, Jintao Fu, Victoria Basile, John S. Corsi, Zeyu Wang, Eric Detsi, in Renewable Energy Vol 155. 189—196 (2020), entitled *Activated alumina as value-added byproduct from the hydrolysis of hierarchical nanoporous aluminum with pure water to generate hydrogen fuel* and by Timothy Lee, Hyeongjun Koh, Alexander K. Ng, Jiaxin Liu, Eric A. Stach, Eric Detsi, in Scripta Materialia 221, 114959—114963 (2022), entitled *Ultrafine nanoporous aluminum by electrolytic dealloying of aluminum-magnesium alloys in glyme-based electrolytes with recovery of sacrificial magnesium*. The authors are thankful to Timothy Lee, Jintao Fu, Victoria Basile, John S. Corsi, Zeyu Wang, Hyeongjun Koh, Alexander K. Ng, Jiaxin Liu, and Eric A. Stach for their discussion and contributions.

References

[1] J. Bitenc, M. Firm, A. Randon Vitanova, R. Dominko, Effect of Cl— and TFSI— anions on dual electrolyte systems in a hybrid $Mg/Li_4Ti_5O_{12}$ battery, Electrochem. Commun. 76 (2017) 29—33, https://doi.org/10.1016/j.elecom.2017.01.013.

[2] F. Cuevas, C. Murray, W. Platzer, A. Heimsath, Large scale solar plants integration in electro-winning copper recuperation process, Energy Proc. 70 (2015) 605—614, https://doi.org/10.1016/j.egypro.2015.02.167.

[3] J.S. Corsi, J. Fu, Z. Wang, T. Lee, A.K. Ng, E. Detsi, Hierarchical bulk nanoporous aluminum for on-site generation of hydrogen by hydrolysis in pure water and combustion of solid fuels, ACS Sustain. Chem. Eng. 7 (2019) 11194—11204, https://doi.org/10.1021/acssuschemeng.9b00481.

[4] J. Petrovic, G. Thomas, Reaction of Aluminum With Water to Produce Hydrogen - 2010 Update, 2011, https://doi.org/10.2172/1219359. United States.

[5] L. Soler, A.M. Candela, J. Macanás, M. Muñoz, J. Casado, Hydrogen generation by aluminum corrosion in seawater promoted by suspensions of aluminum hydroxide, Int. J. Hydrogen Energy 34 (2009) 8511—8518, https://doi.org/10.1016/j.ijhydene.2009.08.008.

[6] H.-T. Teng, T.-Y. Lee, Y.-K. Chen, H.-W. Wang, G. Cao, Effect of $Al(OH)_3$ on the hydrogen generation of aluminum—water system, J. Power Sources 219 (2012) 16—21, https://doi.org/10.1016/j.jpowsour.2012.06.077.

[7] Y.-K. Chen, H.-T. Teng, T.-Y. Lee, H.-W. Wang, Rapid hydrogen generation from aluminum—water system by adjusting water ratio to various aluminum/aluminum hydroxide, Int. J. Energy Environ. Eng. 5 (2014) 87, https://doi.org/10.1007/s40095-014-0087-3.

[8] Y. Yavor, S. Goroshin, J.M. Bergthorson, D.L. Frost, R. Stowe, S. Ringuette, Enhanced hydrogen generation from aluminum—water reactions, Int. J. Hydrogen Energy 38 (2013) 14992—15002, https://doi.org/10.1016/j.ijhydene.2013.09.070.
[9] S.O. Kazantsev, A.S. Lozhkomoev, E.A. Glazkova, I. Gotman, E.Y. Gutmanas, M.I. Lerner, S.G. Psakhie, Preparation of aluminum hydroxide and oxide nanostructures with controllable morphology by wet oxidation of AlN/Al nanoparticles, Mater. Res. Bull. 104 (2018) 97—103, https://doi.org/10.1016/j.materresbull.2018.04.011.
[10] A.I. Osman, J.K. Abu-Dahrieh, D.W. Rooney, S.A. Halawy, M.A. Mohamed, A. Abdelkader, Effect of precursor on the performance of alumina for the dehydration of methanol to dimethyl ether, Appl. Catal. B 127 (2012) 307—315, https://doi.org/10.1016/j.apcatb.2012.08.033.
[11] E. V Kul'ko, A.S. Ivanova, V.Y. Kruglyakov, E.M. Moroz, K.I. Shefer, G.S. Litvak, G.N. Kryukova, Y.Y. Tanashev, V.N. Parmon, Synthesis of aluminum oxides from the products of the rapid thermal decomposition of hydrargillite in a centrifugal flash reactor: II. Structural and textural properties of aluminum hydroxide and oxide obtained from the product of the centrifugal thermal activation of hydrargillite (CTA product), Kinet. Catal. 48 (2007) 316—326, https://doi.org/10.1134/S0023158407020176.
[12] A. Savel'eva, T. Bugrova, A. Petrov, V. Dutov, G. Mamontov, Thermally-activated $Al(OH)_3$: phase transformations and porosity, Key Eng. Mater. 670 (2016) 133—138, https://doi.org/10.4028/www.scientific.net/KEM.670.133.
[13] A.S. Ivanova, Aluminum oxide and systems based on it: properties and applications, Kinet. Catal. 53 (2012) 425—439, https://doi.org/10.1134/S0023158412040039.
[14] A. Pearson, Aluminum oxide (alumina), activated, in: Kirk-Othmer Encyclopedia of Chemical Technology, 2000, https://doi.org/10.1002/0471238961.0103200916050118.a01.
[15] X. Du, X. Su, Y. Wang, J. Li, Thermal decomposition of grinding activated bayerite, Mater. Res. Bull. 44 (2009) 660—665, https://doi.org/10.1016/j.materresbull.2008.06.031.
[16] Z.-Y. Deng, Y.-F. Liu, Y. Tanaka, J. Ye, Y. Sakka, Modification of Al particle surfaces by γ-Al_2O_3 and its effect on the corrosion behavior of Al, J. Am. Ceram. Soc. 88 (2005) 977—979, https://doi.org/10.1111/j.1551-2916.2005.00154.x.
[17] M. Liu, S. Zanna, H. Ardelean, I. Frateur, P. Schmutz, G. Song, A. Atrens, P. Marcus, A first quantitative XPS study of the surface films formed, by exposure to water, on Mg and on the Mg—Al intermetallics: Al_3Mg_2 and $Mg_{17}Al_{12}$, Corros. Sci. 51 (2009) 1115—1127, https://doi.org/10.1016/j.corsci.2009.02.017.
[18] J.T. Kloprogge, L. V Duong, B.J. Wood, R.L. Frost, XPS study of the major minerals in bauxite: gibbsite, bayerite and (pseudo-)boehmite, J. Colloid Interface Sci. 296 (2006) 572—576, https://doi.org/10.1016/j.jcis.2005.09.054.
[19] N. Kumar, K. Biswas, Cryomilling: an environment friendly approach of preparation large quantity ultra refined pure aluminium nanoparticles, J. Mater. Res. Technol. 8 (2019) 63—74, https://doi.org/10.1016/j.jmrt.2017.05.017.
[20] A. Ganguly, S. Sharma, P. Papakonstantinou, J. Hamilton, Probing the thermal deoxygenation of graphene oxide using high-resolution in situ X-ray-based spectroscopies, J. Phys. Chem. C 115 (2011) 17009—17019, https://doi.org/10.1021/jp203741y.
[21] R. Wischert, P. Laurent, C. Copéret, F. Delbecq, P. Sautet, γ-Alumina: the essential and unexpected role of water for the structure, stability, and reactivity of "defect" sites, J. Am. Chem. Soc. 134 (2012) 14430—14449, https://doi.org/10.1021/ja3042383.

[22] W. Cai, J. Yu, M. Jaroniec, Template-free synthesis of hierarchical spindle-like γ-Al_2O_3 materials and their adsorption affinity towards organic and inorganic pollutants in water, J. Mater. Chem. 20 (2010) 4587–4594, https://doi.org/10.1039/B924366F.

[23] A. Srivastav, V.C. Srivastava, Adsorptive desulfurization by activated alumina, J. Hazard Mater. 170 (2009) 1133–1140, https://doi.org/10.1016/j.jhazmat.2009.05.088.

[24] M. Thommes, K. Kaneko, A. V Neimark, J.P. Olivier, F. Rodriguez-Reinoso, J. Rouquerol, K.S.W. Sing, Physisorption of gases, with special reference to the evaluation of surface area and pore size distribution, IUPAC Tech. Rep. 87 (2015) 1051–1069, https://doi.org/10.1515/pac-2014-1117.

[25] O. Mizrahi, N. Amir, E. Pollak, O. Chusid, V. Marks, H. Gottlieb, L. Larush, E. Zinigrad, D. Aurbach, Electrolyte solutions with a wide electrochemical window for rechargeable magnesium batteries, J. Electrochem. Soc. 155 (2008) A103, https://doi.org/10.1149/1.2806175.

[26] H. Yaghoobnejad Asl, J. Fu, H. Kumar, S.S. Welborn, V.B. Shenoy, E. Detsi, In situ dealloying of bulk Mg_2Sn in Mg-ion half cell as an effective route to nanostructured Sn for high performance Mg-ion battery anodes, Chem. Mater. 30 (2018) 1815–1824, https://doi.org/10.1021/acs.chemmater.7b04124.

[27] T. Gao, F. Han, Y. Zhu, L. Suo, C. Luo, K. Xu, C. Wang, Hybrid Mg_2+/Li+ battery with long cycle life and high rate capability, Adv. Energy Mater. 5 (2015), https://doi.org/10.1002/aenm.201401507.

[28] L. Yang, C. Yang, Y. Chen, Z. Pu, Z. Zhang, Y. Jie, X. Zheng, Y. Xiao, S. Jiao, Q. Li, D. Xu, Hybrid $MgCl_2/AlCl_3/Mg(TFSI)_2$ electrolytes in DME enabling high-rate rechargeable Mg batteries, ACS Appl. Mater. Interfaces 13 (2021) 30712–30721, https://doi.org/10.1021/acsami.1c07567.

[29] J. Erlebacher, M.J. Aziz, A. Karma, N. Dimitrov, K. Sieradzki, Evolution of nanoporosity in dealloying, Nature 410 (2001) 450–453, https://doi.org/10.1038/35068529.

[30] C. Soyarslan, S. Bargmann, M. Pradas, J. Weissmüller, 3D stochastic bicontinuous microstructures: generation, topology and elasticity, Acta Mater. 149 (2018) 326–340, https://doi.org/10.1016/j.actamat.2018.01.005.

[31] A.K. Ng, S.S. Welborn, E. Detsi, Time-dependent power law function for the post-dealloying chemical coarsening of nanoporous gold derived using small-angle X-ray scattering, Scr. Mater. 206 (2022) 114215, https://doi.org/10.1016/j.scriptamat.2021.114215.

[32] J.S. Corsi, J. Fu, L. Wang, S.S. Welborn, Z. Wang, E. Detsi, Sacrificial silver recovery during nanoporous gold formation by electrolytic dealloying of gold-silver alloy, J. Electrochem. Soc. 169 (2022) 063501, https://doi.org/10.1149/1945-7111/ac6344.

[33] T. Lee, J. Fu, V. Basile, J.S. Corsi, Z. Wang, E. Detsi, Activated alumina as value-added byproduct from the hydrolysis of hierarchical nanoporous aluminum with pure water to generate hydrogen fuel, Renew. Energy 155 (2020) 189–196, https://doi.org/10.1016/j.renene.2020.03.072.

[34] A. V Bersh, A. V Lisitsyn, A.I. Sorokovikov, M.S. Vlaskin, Y.A. Mazalov, E.I. Shkol'nikov, Study of the processes of steam-hydrogen mixture generation in a reactor for hydrothermal aluminum oxidation for power units, High Temp. 48 (2010) 866–873, https://doi.org/10.1134/S0018151X10060131.

[35] M.S. Vlaskin, A. V Grigorenko, A.Z. Zhuk, A. V Lisitsyn, A.E. Sheindlin, E.I. Shkol'nikov, Synthesis of high-purity α-Al_2O_3 from boehmite obtained by hydrothermal oxidation of aluminum, High Temperature 54 (2016) 322–329, https://doi.org/10.1134/S0018151X16020231.

[36] O. V Al'myasheva, E.N. Korytkova, A. V Maslov, V. V Gusarov, Preparation of nanocrystalline alumina under hydrothermal conditions, Inorg. Mater. 41 (2005) 460–467, https://doi.org/10.1007/s10789-005-0152-7.
[37] F. Maggi, S. Dossi, C. Paravan, L.T. DeLuca, M. Liljedahl, Activated aluminum powders for space propulsion, Powder Technol 270 (2015) 46–52, https://doi.org/10.1016/j.powtec.2014.09.048.

CHAPTER EIGHT

Nanoporous tri-layer of similar elements by etching without sacrificing materials through the Kirkendall effect

8.1 Overview

The major limitation of the selective etching process is that the sacrificial metals etched from the parent alloy, which often account for over 50% of the parent material, usually end up as waste products. Etching without sacrificial material loss is a crucial step toward a sustainable, circular economy. Therefore, in Chapters 8 and 9, we will demonstrate a novel, versatile electrolytic method for making nanoporous metals that nearly eliminates the loss of sacrificial materials. This method, which we call chlorine-neutral etching, uses Cl to introduce nanoporosity in metals through the Kirkendall effect, with nearly complete recovery of Cl. We will showcase this method by making noble nanoporous metal/metal composites, including nanoporous Ag/Ag composites (Chapter 8) and nanoporous Ag/Au composites (Chapter 9), with the expectation that this method could be extended in the future to make nonprecious nanoporous metal/metal composites.

Depletion plating, also known as depletion gilding, has been practiced since ancient civilizations to alter the appearance of metallic objects by depleting Cu near the surface of Au-Cu and Ag-Cu items through selective etching, making them look like pure Au and pure Ag (Sterling Ag), respectively. In modern times, selective etching is widely used to create nanoporous metals, some of which are integral to industries like Raney Ni catalysis. Etching without losing sacrificial materials will enable a circular economy. Here, we demonstrate this new concept by fabricating a trilayer nanoporous Ag | Ag | Ag.

8.1.1 Nanoporous layers and Kirkendall effect

High surface-to-volume ratio nanoporous metals have been intensely studied for applications such as energy storage [1−3], (electro)catalysis [4−7],

Metals as Clean Fuels
ISBN: 978-0-443-13537-8
https://doi.org/10.1016/B978-0-443-13537-8.00006-3

plasmonics [8–10], actuation [11–13], and sensing [14,15]. These nanoporous metals are commonly made by selectively *etching* a sacrificial element from a homogeneous parent alloy using a corrosive medium – a process called "free corrosion dealloying" – which is similar to the depletion gilding process practiced since the Bronze Age in ancient Egypt and Asia to alter the appearance of metallic objects [16,17]. During selective etching, sacrificial elements removed from the parent alloy often become waste products [18–20]. The sample composition of two typical single-phase model parent alloys used to make nanoporous gold and nanoporous silver are $Au_{35}Ag_{65}$ and quenched $Ag_{15}Al_{85}$ at.%, respectively, wherein 65–85 at.% of the parent alloy is sacrificed to create the corresponding nanoporous metal [21,22]. Methods that reduce sacrificial materials or sacrificial chemicals are desirable, as sustainable synthesis methods that enable a circular economy are becoming increasingly crucial. To this end, in prior works, it was demonstrated that the partial recovery of sacrificial materials used to fabricate several nanoporous metals by electrolytic dealloying [23–25]. In the case of nanoporous gold made from a gold-silver alloy, $\approx$35 at.% of the sacrificial silver still could not be recovered [23]. The challenge of developing a selective leaching method that nearly eliminates the loss of sacrificial materials and chemicals remains unresolved.

Previously, Wang et al. reported on a new dealloying method referred to as *reduction-induced decomposition*. Unlike standard dealloying methods that require homogeneous parent alloys, they utilized chemical reduction to selectively etch a sacrificial element (Cl) from a metal chloride compound (AgCl), resulting in a nanoporous metal (NP-Ag) structure [26,27]. However, the Cl in the starting material is irreversibly transferred to the etching solution, ending up as a waste product. A chlorine-neutral etching process with the same material composition before and after etching is desirable. Jin et al. pioneered such a chlorine-neutral process by fabricating a binary structure consisting of a porous Ag layer and a dense Ag layer, starting with a dense Ag substrate and an etching solution containing Cl ions [28]. They electrochemically deposited a layer of AgCl onto this dense Ag substrate used as an electrode (Eq. 8.1), and subsequently reduced the AgCl layer to porous silver (reverse of Eq. 8.1). Despite its chlorine-neutral nature, this process is limited to a near-surface reaction and does not yield a fully porous Ag structure, but rather creates a dense Ag layer and a porous Ag layer. Creating a *fully* porous structure without losing sacrificial materials or sacrificial chemicals remains a challenge. In the present work, we overcome this challenge by incorporating an innovative concept, the Kirkendall effect, to Jin et al.'s processing method, resulting in a *fully* "dealloyed" nanoporous metal structures, thereby demonstrating an intrinsic relationship between the electrochemical growth of AgCl and the Kirkendall effect.

$$Ag + Cl^- \rightleftharpoons AgCl + e^- \quad E_0 = 0.22\ V \text{ versus SHE} \tag{8.1}$$

Initially reported in 1947, the Kirkendall effect is defined as the displacement of an interface A|B between two dissimilar materials, A and B, driven by the difference in the rate at which an atom from A diffuses in B, and an atom from B diffuses in A [29]. This results in the formation of many vacancies in the bulk of the faster diffusing material, which, to achieve equilibrium vacancy concentration, eventually coalesce into pores in the material to minimize the Gibbs free energy of the system [30]. The Kirkendall effect is traditionally associated with pore formation that weakens the mechanical properties of the original structure. However, recent research has embraced the Kirkendall effect in creating nanospheres, nanotubes, and porous structures [31–34]. Our work demonstrates how the Kirkendall effect, combined with electrochemical plating/stripping, forms nanoporous structures from dense (nonporous) materials while sacrificing only less than 2 at. % of the elements involved in the process.

8.2 Experimental methods

8.2.1 Sample preparation

Fabrication of nanoporous Ag supported on a dense Ag foil: A silver shot (purity 3N, *Alfa Aesar*) was rolled to a thick foil (thickness ≈30–50 μm) using a tabletop hand rolling mill. The resulting foil was cut into several strips with dimensions of 5 × 30 mm and used as electrodes in an electrochemical cell for the fabrication of a substrate-supported NP-Ag (see Fig. 8.1a). In doing so, one of the Ag strips served as the working electrode and another as the counter electrode in a three-electrode symmetric float cell. An Ag wire was utilized as the reference electrode in a 0.1 M NaCl aqueous solution used as the electrolyte. To maintain a suitable interelectrode distance, it was ensured that the distance between working and counter electrodes remained below 25 mm. Cyclic voltammetry (CV) measurements were performed using a Bio-Logic SP-300 multichannel potentiostat/galvanostat. A forward oxidation scan from −0.35 to 0.35 V versus Ag/AgCl (scan rate of 0.1 mV/second) develops a layer of AgCl at the surface of the electrode. The reverse reduction scan from 0.35 V to −0.35 V versus Ag/AgCl (scan rate

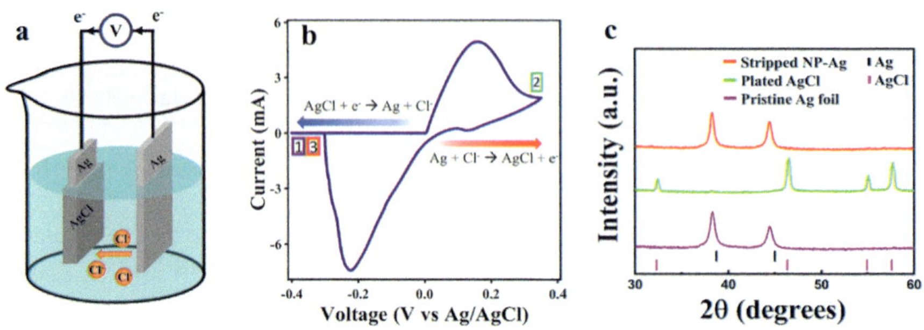

Figure 8.1 (a) Schematic of the symmetric float cell used in our experiments, with Ag as both the working and counter electrodes in a Cl ion containing electrolyte (b) characteristic cyclic voltammetry (CV) of our plating-stripping reaction on Ag foil, where the forward red arrow represents the electrochemical growing of AgCl, and the reverse blue arrow represents the electrochemical reduction of AgCl to Ag. (c) X-ray diffraction (XRD) measurements taken at four states during the CV – the pristine state (1, in purple), the AgCl-plated state (2, in green), and after AgCl reduction on a thin Ag substrate forming NP-Ag (3, in red).

0.1 mV/second) returns the AgCl to its Ag state. Sample characterization was carried out at various stages during the CV cycle. For surface characterization by SEM, a piece of Scotch tape was used to delaminate a region of the top AgCl layer to reveal the *Kirkendall effect-induced nanoporous* Ag (KE-NP-Ag) underneath. Imaging was performed at the "interface" created by the delamination, such that within one image, we can see the nature of the sample before delamination along with the delaminated region. The same cell configuration was used for all data shown in this report unless otherwise specified.

Fabrication of a freestanding tri-layer nanoporous Ag | Ag | Ag film from a thin Ag foil: To create a freestanding tri-layer nanoporous Ag | Ag | Ag film (NP Ag | Ag | Ag), we reduce the thickness of the foil to within the penetration depth of the Kirkendall effect. A rough estimation for this penetration depth is measured from the PFIB/SEM images seen in Fig. 8.2, where it appears that the porosity extends 12 μm below the surface. For this report, an Ag leaf with a thickness of approximately 2 μm is used, as it is below the thickness threshold. The same electrochemical setup configuration for the electrochemical plating and stripping on a thick film is used (see Fig. 8.1a), except that for the working electrode, pure genuine Ag leaf (heavy Ag foil, *Brambilla*) with thickness ≈2 μm replaced the 30–50 μm-thick Ag foil of the previous section. The leaf was cut into several strips with dimensions of 5 × 30 mm. Aside from the working electrode, all other aspects of the cell configuration remained the same as in Section 8.2.1. To investigate

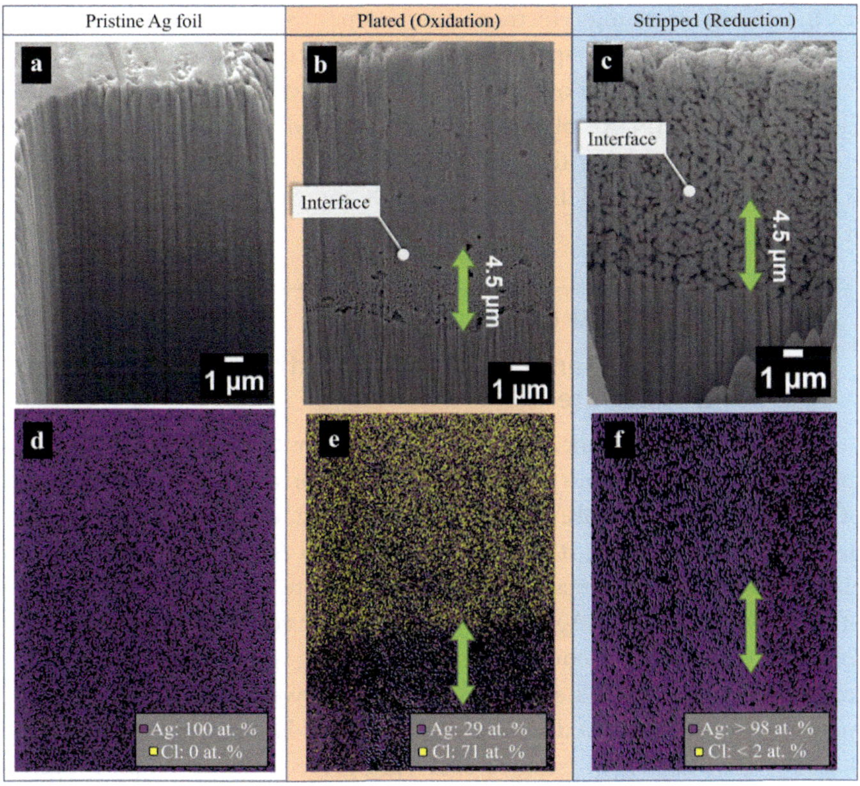

Figure 8.2 PFIB/SEM images of (a) the pristine thick Ag foil, (b) the Ag foil after AgCl is plated on the surface, with a nanoporous structure revealed underneath, and (c) the Ag foil after the AgCl layer has been reduced back to Ag, resulting in two NP-Ag layers atop an Ag substrate. (d–f) Are the corresponding EDS maps and chemical compositions for the PFIB/SEM images. Green arrows highlight the distance between the KE-NP-Ag | r-NP-Ag interface and the underlying bulk Ag.

the influence of temperature on the etching process, a series of experiments were conducted with the reaction vessel immersed in an ice-water bath (temperature measured at 0–3°C). This arrangement ensured that the electrolyte level within the reaction vessel remained lower than the level of the surrounding ice-water bath.

8.2.2 Materials characterization

X-ray diffraction (XRD) patterns were collected using a Rigaku Miniflex powder diffractometer (Cu X-ray source, operating at 40 kV/15 mA) with a 0.02 degrees step size and a 10 degrees/min scanning speed. A

JEOL 7500F scanning electron microscope (SEM) equipped with an energy-dispersive X-ray spectroscopy (EDS) detector was used to characterize the microstructure of the etched Ag at various stages of NP-Ag fabrication. A TESCAN Xe plasma-FIB (PFIB)/SEM was used to characterize cross- sections of thicker foils. Small-angle X-ray scattering was performed at the Dual Source and Environmental X-ray Scattering (DEXS) facility at UPenn.

8.3 Results and discussion

8.3.1 Electrochemical plating and stripping of AgCl on a substrate

To enable the Kirkendall effect, we deposit AgCl onto a dense Ag substrate from an aqueous solution free of Ag ions. This creates the system of two dissimilar materials, AgCl, and metallic Ag, necessary for the Kirkendall effect to take effect. It is emphasized that no Ag ions are present in the electrolyte solution; the starting dense Ag substrate provides the only source of Ag needed to deposit AgCl. Using a symmetric cell configuration with Ag foils as the working and counter electrodes, an Ag wire as the reference electrode, and a 0.1 M NaCl electrolyte (Fig. 8.1a), we perform cyclic voltammetry (CV) to deposit and reduce AgCl. An oxidation scan from −0.35 to 0.35 V versus Ag/AgCl (scan rate of 0.1 mV/second) develops a layer of AgCl at the surface of the electrode. The reverse reduction scan from 0.35 V to −0.35 V versus Ag/AgCl (scan rate 0.1 mV/second) returns the AgCl to its Ag state.

During the forward scan (AgCl deposition), Cl ions from the electrolyte react with the Ag foil working electrode, following Eq. (8.1), to form a coating of AgCl on the Ag foil substrate. Conversely, the reverse scan reduces the AgCl layer, revealing NP-Ag. We use an Ag wire as our reference electrode, which, when employed in a 0.1 M NaCl electrolyte, forms an insoluble surface layer of AgCl, creating an Ag/AgCl wire. Consequently, the CVs performed in this study use a reference electrode considered equivalent to an Ag/AgCl reference electrode stored in 0.1 M Cl ion solution (0.29 V vs. SHE).

The CV signature in Fig. 8.1b shows distinguished oxidation and reduction peaks during the plating and reduction of AgCl onto a dense, pure Ag substrate, where plating and reduction reactions are indicated by red and blue arrows, respectively. This reaction differs from typical plating-stripping reactions in two key aspects: (i) The plating reaction occurs during oxidation in the Ag/AgCl system, while most common plating reactions occur from reducing metal ions to their metal state. (ii) This Ag/AgCl system is surface-limited, as depicted by the distinguished oxidation peak (Fig. 8.1b), unlike the "unlimited" plating seen in most plating/stripping systems [35,36].

The substrate surface was analyzed using X-ray diffraction (XRD) at the following three states in the CV scan, highlighted in Fig. 8.1b: pristine Ag foil (state 1, purple box), after plating AgCl (state 2, green box), and after reduction of AgCl (state 3, red box). The XRD pattern (Fig. 8.1c) of the pristine foil matches that of pure Ag (purple; ICSD Collection Code #22434) [37]. After plating with AgCl, peaks matching that of AgCl (green; ICSD Collection Code #56538) are introduced along with peaks of the substrate [37]. After reducing AgCl, the XRD peaks return to pure Ag (red). Later, we will provide further characterization of this system.

The following sections hold further characterization of two relevant systems: (i) substrate-supported NP-Ag, (ii) and fully nanoporous freestanding trilayer NP Ag | Ag | Ag films, to demonstrate how this electrochemical plating/stripping method can be used to fabricate fully nanoporous films from bulk metals.

8.3.2 Partially porous structures by electrochemical plating and stripping of AgCl on thick Ag foil

We start by creating a partially "dealloyed" NP-Ag system consisting of a thick NP-Ag layer (thickness ≈ 12 μm) on a dense (nonporous) Ag support, to elucidate the mechanism of introducing porosity through plating/stripping. Dense Ag foil strips (rolled to a thickness $\approx 30-50$ μm) were used as the working electrode and counter electrode in our symmetric cell setup. A single CV scan is performed at room-temperature to plate AgCl (forward CV scan) on the dense Ag foil electrode and reduce the plated AgCl to Ag (reverse CV scan). Both oxidized (or plated) and reduced (or stripped) samples are characterized using XRD, small-angle X-ray scattering (SAXS), SEM, electron-dispersive X-ray spectroscopy (EDX) and focused ion beam techniques.

We used a plasma focused-ion beam (PFIB) SEM to investigate the cross sections of the thick Ag foil in its pristine, oxidized, and reduced states. Representative PFIB/SEM data are compiled in Fig. 8.2. As expected, the cross-section of the pristine Ag foil (Fig. 8.2a) reveals no distinct microstructural features, and no chemical elements other than Ag, confirmed by energy dispersive X-ray spectroscopy (EDS) elemental analysis mapping (Fig. 8.2d, purple region). Unexpectedly, the cross-sectional image of the AgCl-plated sample reveals obvious nanoporous structures (green arrow, Fig. 8.2b and e) extending below the surface of the Ag substrate, underneath the AgCl layer (Cl in yellow, and Ag in purple, Fig. 8.2e).

The presence of a nanoporous Ag layer underneath the plated AgCl layer indicates that the mechanism for AgCl layer formation is more complex than typical plating. We attribute this nanoporous structure to the Kirkendall effect, through which Ag atoms are extracted from the substrate to create a thick layer of AgCl at the surface. Nanoporosity formation in the Ag substrate via the Kirkendall effect involves the diffusion of Ag and Cl atoms in opposite directions and at different rates: (i) a relatively fast diffusion of Ag atoms from the bulk of the Ag substrate toward the interface, where they react with Cl from the electrolyte to form an AgCl layer, coupled with (ii) a relatively slow diffusion of Cl from AgCl layer toward the Ag substrate. The higher diffusion rate of Ag equates to more Ag being removed from the Ag substrate than Cl are introduced, leaving vacancy clusters which self-organize into nanoporous structures (seen below the AgCl layer in Fig. 8.2b) to minimize the energy of the system.

In Fig. 8.2c, two distinct layers of nanoporous Ag ($\approx$12 μm total thickness) are visible at the substrate's surface. Both Fig. 8.2b and c have an interface separating the nanoporous Ag layer formed by Kirkendall effect and the nanoporous Ag layer from reduction. This interface is blurry in Fig. 8.2c due to focused ion beam milling, but a closer inspection reveals the interface. In both Fig. 8.2b and c, the green arrows demarcating the distance between this interface and underlying bulk Ag layer are consistent with each other, measuring 4.5 μm. This result combines the new findings introduced here regarding the Kirkendall effect with those of Jin et al. The topmost layer of nanoporous Ag is from reducing AgCl back to its porous structure, which we will henceforth refer to as *reduction-induced nanoporous Ag* (r-NP-Ag). The second layer of porous Ag stems from atomic diffusion to the surface, which we will refer to as *Kirkendall effect-induced nanoporous Ag* (KE-NP-Ag).

We used SEM and EDX to further examine the morphology and chemical composition of the deposited r-NP-Ag and the underlying KE-NP-Ag in Fig. 8.3. For surface characterization by SEM, a piece of Scotch tape was used to delaminate a region of the top AgCl layer to reveal the KE-NP-Ag underneath. Imaging was performed at the "interface" created by the delamination, such that within one image, we can see the nature of the sample before delamination (either AgCl or r-NP-Ag) along with the delaminated

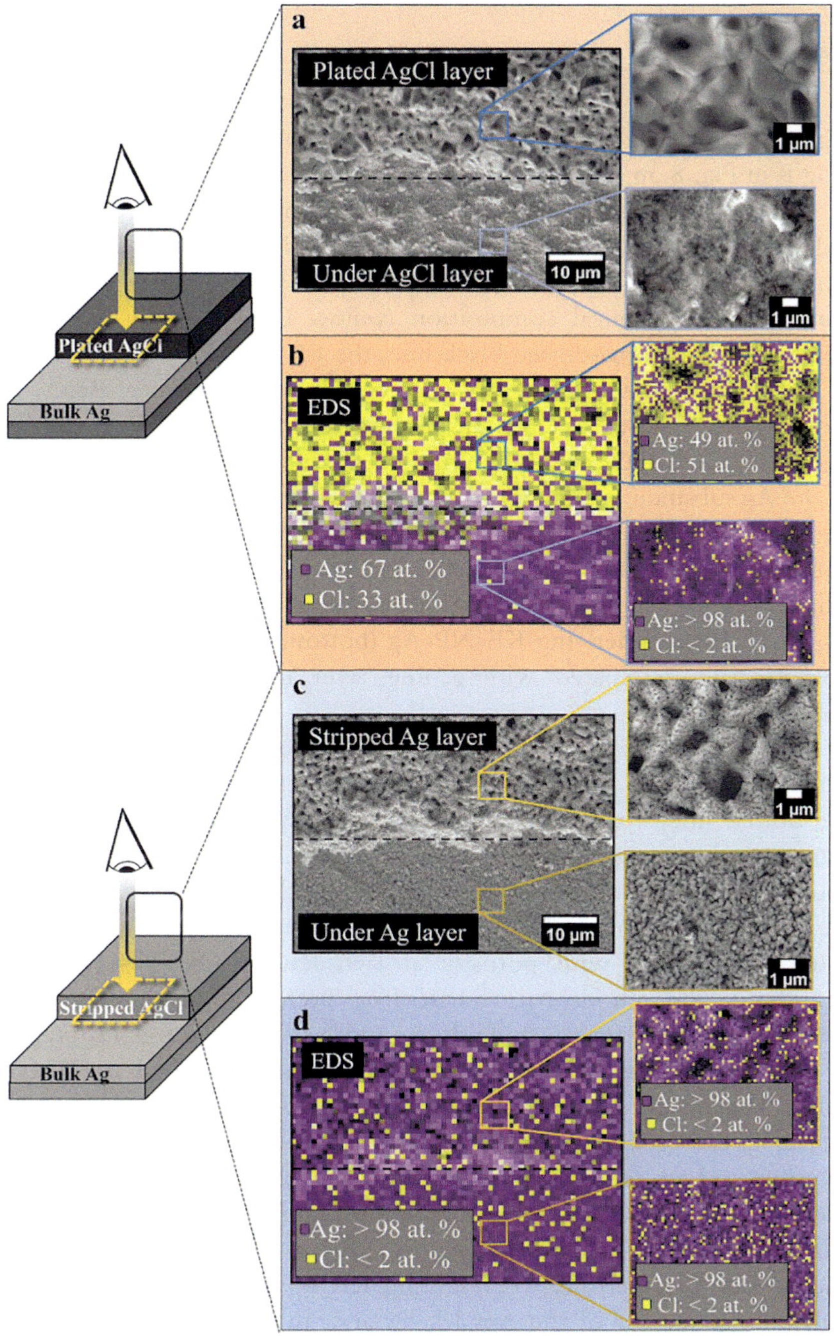

Figure 8.3 SEM images of the top-down view of the plated and reduced thick Ag foil. As illustrated by the schematics on the left, (a and c) show the sample when viewed normal to the surface, after a portion of the top layer (AgCl-plated or reduced Ag) has been peeled off (delaminated) using Scotch tape. (a) Depicts the interface of a plated sample where delamination occurred, with the AgCl layer intact above the dotted line, and the AgCl layer removed below the dotted line. (c) Depicts the interface where the r-NP-Ag layer is delaminated from the substrate. The figure insets show higher magnification images of the top layer (AgCl or r-NP-Ag) and underneath the top layer, respectively. (b) Shows the corresponding EDS maps, confirming the composition of the top layer to be AgCl after plating, and the layer underneath it to be almost pure Ag. (d) Shows the EDS maps, after the stripping of AgCl back to Ag, and confirms that the entire structure has returned to Ag.

region (KE-NP-Ag) below it (Fig. 8.3a and c). Accompanying EDS mapping is in Fig. 8.3b and d, respectively.

Fig. 8.3a focuses on a partially delaminated region of the AgCl plated foil. Both the top AgCl layer (above the dotted black line) and the underlying KE-NP-Ag (below the black dotted line) are visible. EDS maps confirm their chemical composition (yellow for Cl; purple for Ag in Fig. 8.3b). Above the dotted black line, the EDS gives an Ag-to-Cl ratio of $\approx 1{:}1$, consistent with the presence of AgCl at the surface. Below the dotted black line, the EDS indicates almost pure Ag (>98 at. %), with only <2% residual Cl, indicating minimal Cl diffusion from the AgCl layer to the Ag substrate, in agreement with the Kirkendall effect. The insets of Fig. 8.3a and b show selected regions (blue box) for further analysis at higher magnification, revealing a stark difference in structural morphology. The plated AgCl layer consists of >1 μm island-like aggregates (top inset, Fig. 8.3a). The underlying KE-NP-Ag (bottom inset, Fig. 8.3a) shows a porous structure coated with a fine layer of AgCl, which resisted delamination.

Fig. 8.3c focuses on a region of the foil after stripping, where the r-NP-Ag (above the dotted line) and the underlying KE-NP-Ag (below the dotted line) are both visible, displaying the structural morphology shown in the insets. The EDS map in Fig. 8.3d shows nearly pure Ag above and below the dotted line, indicating the reduction of AgCl to Ag. The reduced surface layer, r-NP-Ag, displays a hierarchical nanoporous morphology, with porosity at two length scales, ≈ 1 μm and ≈ 100 nm. The bimodal nature arises from preferential percolation pathways [38,39] that develop along the boundaries between AgCl aggregates to allow for facile electrolyte transport throughout the thick AgCl layer, and eventually form the micron-scale channels seen in the top inset of Fig. 8.3c. As AgCl aggregates are reduced, Cl ions return to the etching solution, and the remaining Ag atoms rearrange into a nanoporous structure (top inset in Fig. 8.3c). The bottom inset in Fig. 8.3c shows a different structural morphology than the top inset, with larger pore sizes and no bimodal porosity, representing KE-NP-Ag, which arose in the bulk during plating of the AgCl layer and, therefore, is not subject to the development of percolation pathways. We attribute the difference in morphology in the bottom insets of Fig. 8.3a and c (both representing KE-NP-Ag) to the thin layer of AgCl that resists delamination, which may hide some of the nanostructures.

8.3.3 Fully porous structures by electrochemical plating and stripping of AgCl on thin Ag foils

Thus far, we have developed a better understanding of how our electrochemical techniques create partially "dealloyed" nanoporous metals (i.e., nanoporous metals on a supporting dense layer). Now, we will show that by reducing the thickness of the supporting foil to within the penetration depth of the Kirkendall effect, *fully* "dealloyed" or freestanding nanoporous structures can be created. Furthermore, AgCl plating/reduction occurs on both sides of the starting substrate, which we expect to form a tri-layer nanoporous Ag | Ag | Ag. A rough estimation for this penetration depth is measured from the PFIB/SEM images seen in Fig. 8.2, where it appears that the porosity extends 12 μm below the surface. Here, we aim to prove our theory by performing plating/stripping experiments a ≈2 μm-thick pure Ag leaf, (heavy Ag foil, *Brambilla*) in contrast to the 30–50 μm-thick foil used in the previous section.

8.3.3.1 Proof-of-concept: Tri-layer nanoporous Ag | Ag | Ag by electrochemical plating and stripping on a thin Ag substrate

Here, we demonstrate that our method can produce *fully* nanoporous metals from a dense metal without the loss of sacrificial materials. As noted earlier, etching methods remove the more reactive metal from a parent alloy, leaving behind a nanoporous structure of the more noble metal (for example, etching Al from an AgAl alloy to form nanoporous silver - NP-Ag). However, in our investigation we will demonstrate that sacrificing Al (or other metals) is no longer necessary to make NP-Ag. Through electrochemical plating and stripping of silver foil, we can create NP-Ag.

We conducted the following electrochemical plating/stripping reactions at a relatively low temperature (around 0–3°C) using a ≈2 μm pure Ag foil. The typical CV associated with this process is shown in Fig. 8.1b. Originally, these experiments were performed at room temperature (discussed in the next section; Section 8.3.3.2), and 0–3°C (discussed here) to demonstrate how the morphology of the fabricated nanoporous metals can be impacted by the reaction temperature, as thermal coarsening is widely known to manipulate pore sizes in nanoporous metals [40–42]. Because dealloying is a surface diffusion-limited process, applying heat allows for easier surface diffusion, resulting in larger ligaments and pores. Conversely, dealloying at a colder temperature can slow surface diffusion, resulting in smaller ligaments and pores [43,44]. Unfortunately, we found that during NP-Ag fabrication, room-temperature electrochemical plating/stripping yields pore sizes exceeding the measurement range of our SAXS instrument. As such, we focus on the experiments performed at 0–3°C.

Figs. 8.4 and 8.5 show the typical SEM and EDS data associated with the plating and reduction of AgCl on this thin Ag film at 0–3°C. SEM and EDS for the AgCl-coated sample top surface (Fig. 8.4a and b) and the cross-section (Fig. 8.4c and d) are compiled in the orange box. SEM and EDS for the reduced sample top surface (Fig. 8.5a and b) and cross-section (Fig. 8.5c and d) are compiled in the blue box. Plating yields a trilayer thin film – with two layers of AgCl aggregates sandwiching a layer of KE-NP-Ag (Fig. 8.4c). The corresponding EDS map shows the pores in the KE-NP-Ag have been partially filled with AgCl (Fig. 8.4d), making it difficult to discern the interface between the layers. However, we see in Fig. 8.5c that after stripping we can identify clear interfaces, demarcating three layers – one layer of KE-NP-Ag sandwiched between two layers of r-NP-Ag. Magnified versions of Figs. 8.4c and 8.5c of these images are

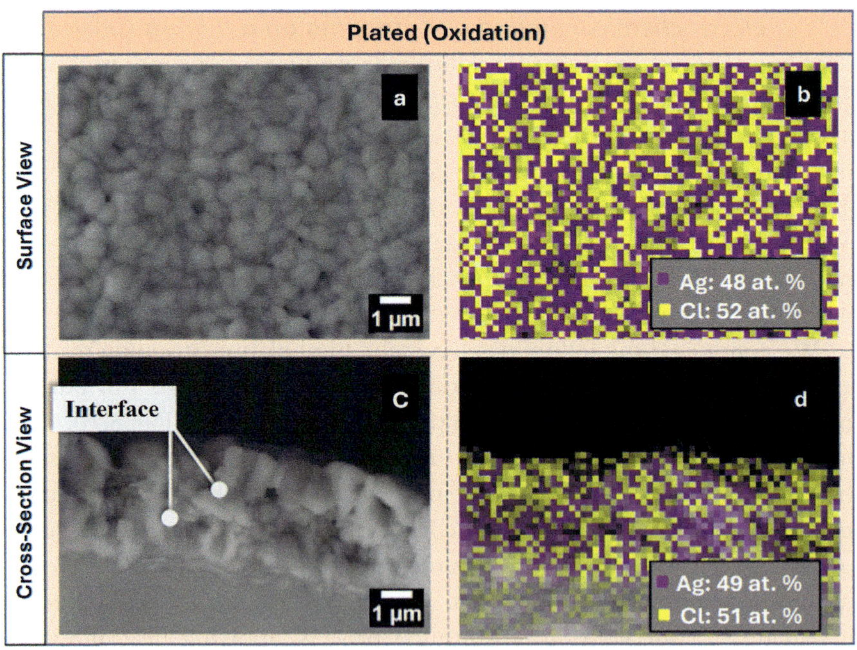

Figure 8.4 SEM and EDS characterization for the electrochemical growing and reduction of AgCl in an ice-water bath (0–3°C) on a thin Ag film. Highlighted in orange "plated oxidation" are the AgCl-plated (a) surface and (c) cross-sectional view. Notice the interface separating the trilayered structure in (c). (b and d) Show corresponding EDS maps with compositional analysis.

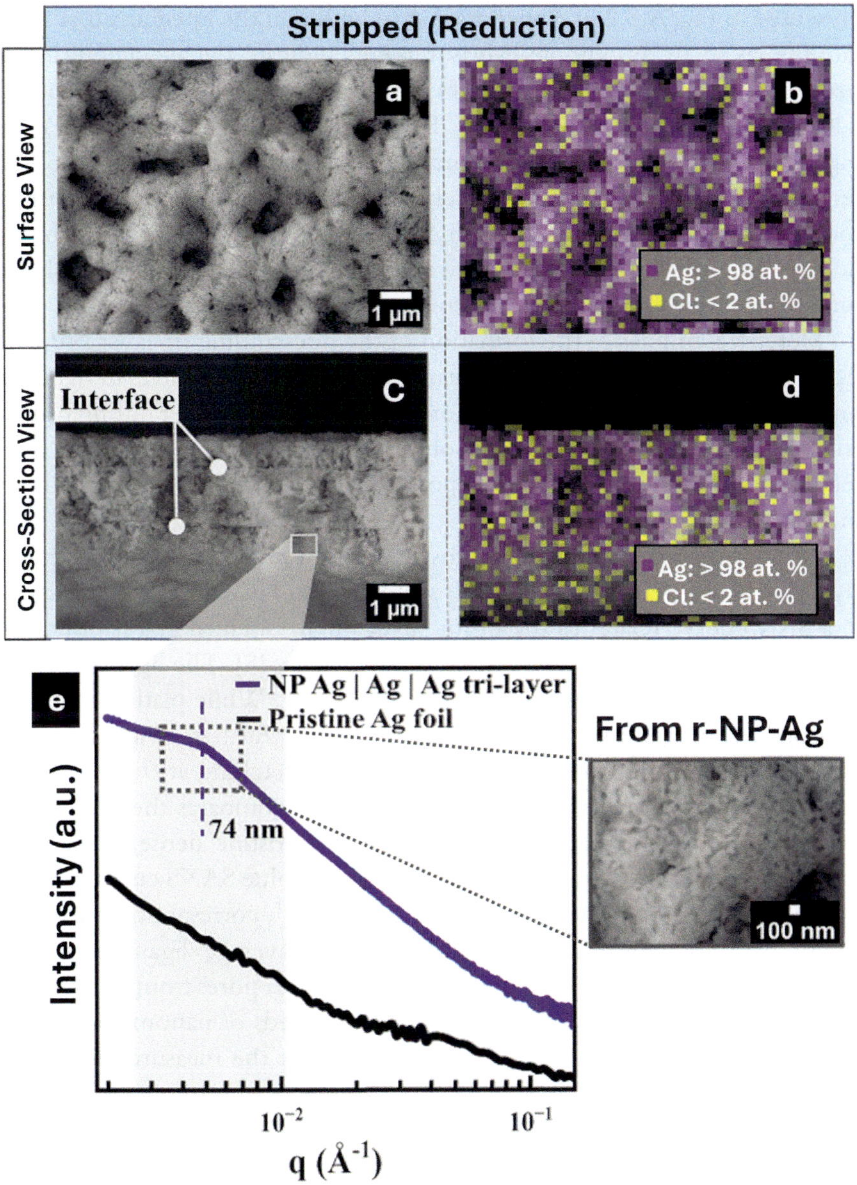

Figure 8.5 SEM and EDS characterization for the electrochemical growing and reduction of AgCl in an ice-water bath (0–3°C) on a thin Ag film. Highlighted in blue 'stripped reduction' are the r-NP-Ag (a) surface and (c) cross-sectional view. Notice the interface separating the trilayered structure in (c). (b and d) Show corresponding EDS maps with compositional analysis. (e) Small-angle X-ray scattering (SAXS) curves demonstrating the formation of bimodal nanoscale porosity after chlorine-neutral etching of a thin Ag foil at 0–3°C. The curve of the pristine Ag foil (in black) shows no scattering peak, indicating no porosity. The curve of the etched sample feature one dominant peak, with the peak representing the outer layers of r-NP-Ag measuring ≈74 nm. A representative high magnification SEM image of this r-NP-Ag is shown in the inset of the SAXS plot.

presented in Fig. A.8.2 of Appendix 8.A to highlight the interfaces that separate the layers in this sample. Measurements indicate the average ligament size of the KE-NP-Ag is ~180 nm, and the average ligament size of the r-NP-Ag is ~80 nm (shown in Fig. A.8.1a and b, respectively). In Fig. 8.5, a bimodal nanoporous structure is seen in the surface view of the reduced sample (contrary to the lack of hierarchical structure seen during room-temperature etching, seen in Section 8.3.3.2). We attribute this to slower diffusion kinetics during lower-temperature etching, which allows time for the development of preferred percolation pathways.

Here, we emphasize the formation of the freestanding tri-layer NP Ag | Ag | Ag *with minimal loss of sacrificial material.* The process involves drawing Cl ions from a Cl-ion containing solution to form the AgCl layer during plating and returning the Cl ions to the solution in a *chlorine-neutral reaction* after reduction, resulting in a fully nanoporous structure without sacrificing any material.

SAXS confirms the holistic porosity throughout the nanoporous metal layers, as it can penetrate the entire sample [45]. The peak position (q_{max}) in a SAXS curve relates to the ligament-ligament distance in a nanoporous metal through the approximation in Eq. (8.2) [46–48]. The ligament size is approximately half the ligament-ligament distance. While plating/stripping at room temperature resulted in trilayer NP Ag | Ag | Ag with pore sizes beyond the measurement range (see Fig. A.8.3), etching at 0–3°C brings the pore size within measurable range. Fig. 8.5e compares the SAXS plot of the 0–3°C r-NP-Ag (shown in blue) to a pristine dense, nonporous foil (shown in black). There is a visible peak in the blue SAXS curve (marked by a blue dashed line), with the q_{max} at 0.0052 Å^{-1}, corresponding to a ligament size of 74 nm, in agreement with the average ligament size of ≈80 nm from the SEM image in Fig. A.8.1b. Larger pores from the Kirkendall effect and percolation, in the range of hundreds of nanometers to microns, were expected (Fig. 8.5a) but were outside the measurement range of the instrument and not visible in the data.

$$q_{\max} \approx 1.23\frac{2\pi}{d} \tag{8.2}$$

8.3.3.2 *Effect of temperature on the morphology of the tri-layer NP Ag | Ag | Ag during electrochemical plating and stripping*

Below, we show the effect of room-temperature electrochemical plating/stripping on a thin silver substrate (≈2 μm silver leaf), as opposed to the electrochemical plating/stripping at 0–3°C that is discussed in Section 8.3.3.1. We aim to demonstrate how the morphology of the fabricated nanoporous metals can be impacted by the reaction temperature. Figs. 8.6 and 8.7 show the results of plating and reducing AgCl on this thin film performed at room temperature. SEM and EDS for the AgCl-coated sample top surface (Fig. 8.6a and b) and the cross-section (Fig. 8.6c and d) are compiled in the orange box. SEM and EDS for the reduced sample surface (Fig. 8.7a and b) and the cross-section (Fig. 8.7c and d) are compiled in the blue box.

After plating AgCl on the surface of the Ag leaf, there is a distinct interface between the AgCl layer and the KE-NP-Ag, thus separating the Ag leaf into three layers – one KE-NP-Ag layer sandwiched between two layers of AgCl (Fig. 8.6c). Magnified versions of Figs. 8.6c and 8.7c of these images are presented in Fig. A.8.4 of Appendix 8.A to highlight the interfaces that separate the layers in this sample. The two AgCl layers consist of dense

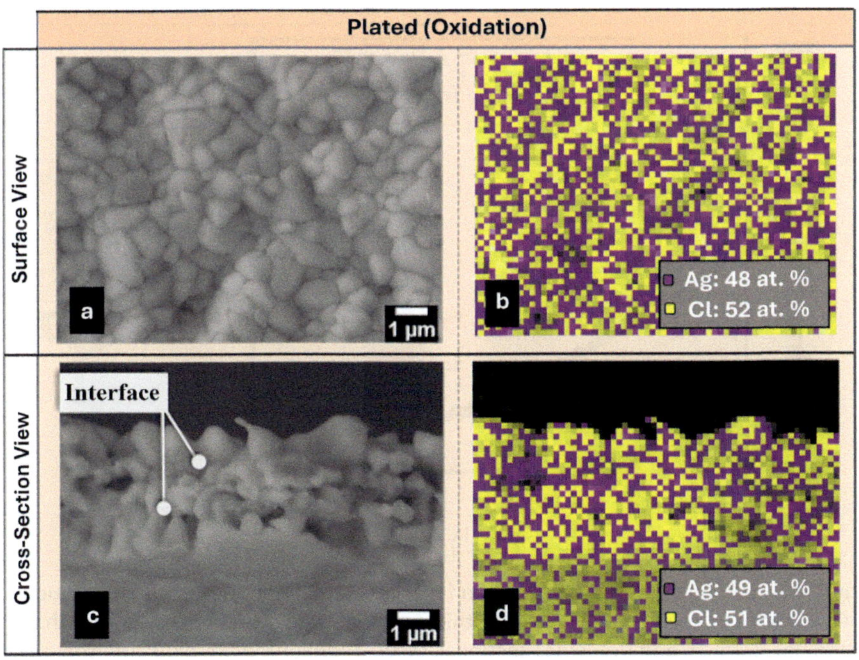

Figure 8.6 SEM and EDS characterization for the electrochemical growing and reduction of AgCl at room temperature (25°C) on a thin Ag film. Highlighted in orange are the AgCl-plated (a) surface and (c) cross-sectional view. Notice the interface separating the trilayered structure in (c). (b and d), Show corresponding EDS maps with compositional analysis.

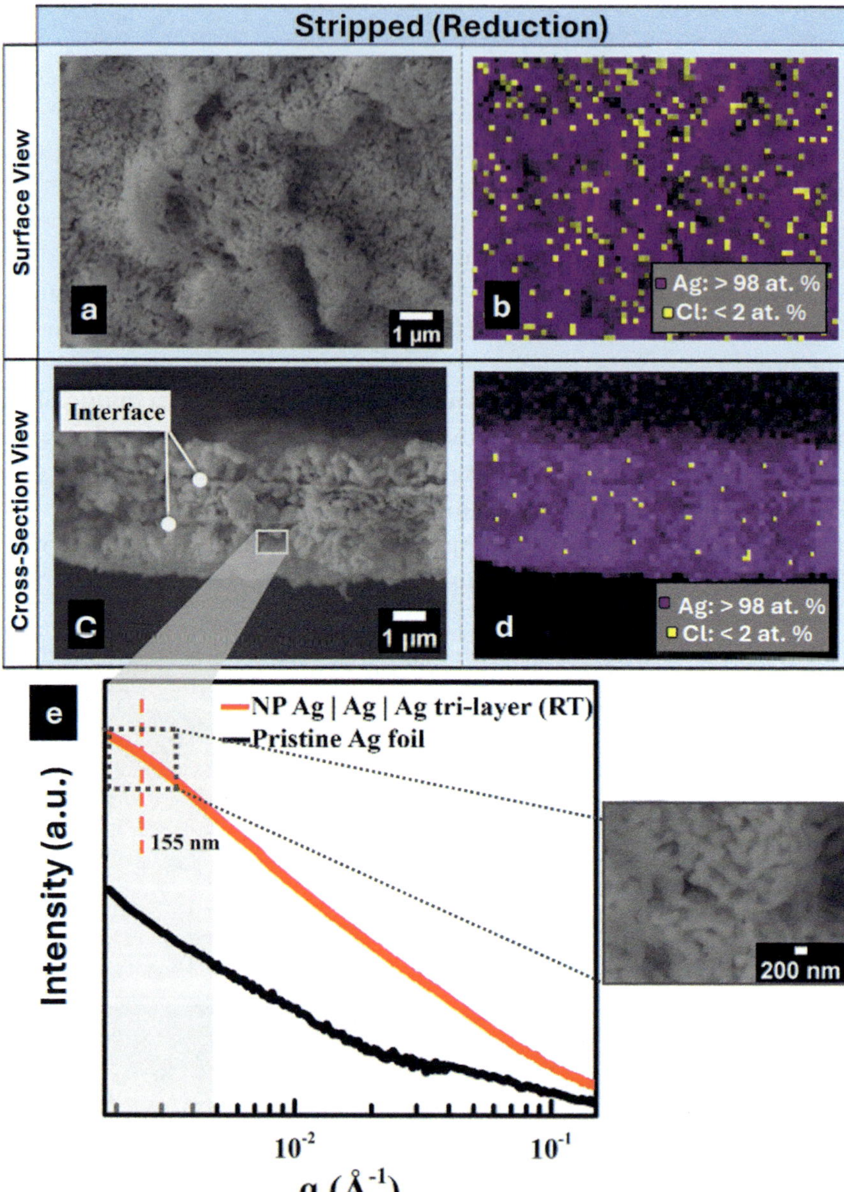

Figure 8.7 SEM and EDS characterization for the electrochemical growth and reduction of AgCl at room temperature (25°C) on a thin Ag film. Highlighted in orange are the AgCl-plated (a) surface and (e) cross-sectional view. Highlighted in blue are the r-NP-Ag (a) surface and (c) cross-sectional view. Notice the interface separating the tri-layered structure in (c). (b and d) Show corresponding EDS maps with compositional analysis. (e) Small-angle X-ray scattering (SAXS) curves demonstrating the formation of nanoscale porosity after chlorine-neutral etching of a thin Ag foil at room temperature. The curve of the pristine Ag foil (in black) shows no scattering peak, indicating no porosity. The curve of the etched sample features one dominant peak, with the peak representing the outer layers of r-NP-Ag measuring $\approx$155 nm. A representative high-magnification SEM image of this r-NP-Ag is shown in the inset of the SAXS plot.

aggregates, while the KE-NP-Ag middle layer has a porous structure. The EDS map in Fig. 8.6d shows AgCl throughout the entire cross-section, indicated by the 1:1 ratio of Ag:Cl. We attribute this to the facile transport of electrolyte throughout the full thickness of the leaf, allowing for AgCl plating to occur within the pores of the KE-NP-Ag, as well. Like the result of plating on a thick Ag foil, SEM imaging normal to the surface, in Fig. 8.6a, reveals large aggregates of AgCl covering the surface of the thin Ag leaf. The reduction of the AgCl to Ag results in the freestanding trilayer NP-Ag structure seen in Fig. 8.7c. Again, the interfaces demarcate three layers — one KE-NP-Ag layer sandwiched between two layers of r-NP-Ag, as is confirmed by the EDS map in Fig. 8.7d. The KE-NP-Ag has an average ligament size of ≈ 300 nm (Fig. A.8.3a), while the r-NP-Ag has an average ligament size of ≈ 148 nm (Fig. A.8.3b).

Contrary to the hierarchical bimodal structures observed after the reduction of AgCl on a thick Ag foil and when done in an ice-water bath, the surface view of the r-NP-Ag on a thin foil (Fig. 8.7a) reveals no bimodal porosity. The thinness of the AgCl layer and the room-temperature etching conditions lead to easier diffusion of the Ag atoms. Similar to the mechanism in typical free corrosion dealloying of nanoporous gold, at room temperature, the propagation of the dealloying front is sufficiently fast along the grains in the AgCl layer such that percolation pathways need not form, resulting in a lack of bimodal structure [49]. We still witness the formation of bimodal structures after cold dealloying, because, unlike at room temperature, we expect the rate of diffusion of the dealloying front to be suppressed. Furthermore, the pore and ligament size of the r-NP-Ag is also coarser after room-temperature plating/stripping, as we can see from the high magnification images in Appendix 8.A, comparing Fig. A.8.3 (room-temperature) to Fig. A.8.1 (0–3°C).

Again, SAXS was used to determine the characteristic length scale in this nanoporous structure. However, as we mentioned earlier, the structure size of room-temperature sample is outside the measurement range of our SAXS instrument. We expect to see SAXS peaks representing pores of ≈ 150 nm. As we see from Fig. 8.7e, most of the peaks have been cut off by the lower measurement limits. In Appendix 8.A, we present a plot (Fig. A.8.5) comparing the SAXS data from this electrochemical plating and stripping at room temperature and in an ice-water bath. Aligning with our current discussion, the cold dealloyed NP Ag | Ag | Ag displays a SAXS peak at higher q-values, indicating smaller pores throughout the film.

8.3.4 Model for NP metal formation via electrochemical plating and stripping in combination with the Kirkendall effect

The formation of the NP metals happens through a two-step process: (1) *Oxidation* of a dense Ag foil will grow AgCl at the surface through a reaction between Ag atoms and Cl ions in solution, seen in Fig. 8.8a. AgCl particles nucleate at the substrate surface, and Ag atoms are drawn from the bulk via the Kirkendall effect to supply reactants for continued growth of these AgCl particles, as shown in Fig. 8.8b. Eventually, the substrate surface is coated in large micron-scale particles, with the voids formed via the Kirkendall effect aggregating to form a nanoporous network immediately below the surface of the substrate, as shown in Fig. 8.8c. (2) *Reduction* of the plated AgCl layer

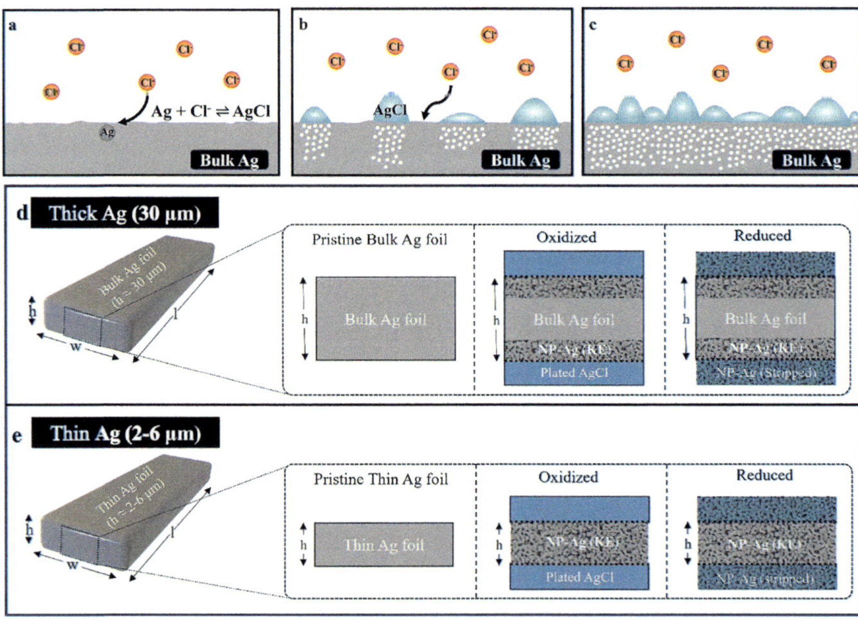

Figure 8.8 (a–c) Schematic of the proposed mechanism of electrochemical plating-stripping: (a) Initiation and nucleation of AgCl growth at the surface of an Ag substrate. (b) Growth of AgCl islands at the surface requires Ag atoms from the bulk, which diffuse toward the surface via the Kirkendall effect, leaving pores in the bulk. (c) Growth at the surface stops when it is fully covered, leaving a porous network within the Ag bulk. (d) Shows the steps within this mechanism applied to a thick Ag film. The rightmost image illustrates the resulting structure of NP-Ag layers with a bulk Ag support at the center. (e) Shows the steps within this mechanism applied to a thin Ag leaf. The rightmost image illustrates the resulting structure of a free-standing trilayered NP-Ag without any supporting structure.

is the reverse of the AgCl layer growth. Reducing AgCl back to Ag creates a nanoporous structure, whose morphology is regulated by diffusion kinetics. Although not intensively studied, we concluded that if diffusion is slow, the reduction of the AgCl layer will form preferential percolation pathways along the grains between AgCl aggregate, to facilitate electrolyte access through the material, which results in bimodal nanoporous structures, seen in Figs. 8.3 and 8.5 [27].

We demonstrate this electrochemical plating/stripping on various Ag-containing substrates. Illustrated in Fig. 8.8d, an AgCl layer is grown on both sides of a thick Ag substrate, with the KE-NP-Ag extending into the bulk. Upon reduction, the AgCl also becomes a nanoporous structure, resulting in several layers of NP-Ag, supported by the bulk substrate in the center (rightmost image, Fig. 8.8d).

However, it is often more useful to have a fully "dealloyed" or free-standing nanoporous structure without a supporting structure, and we have accomplished this using thin Ag-containing substrates as the starting material (Fig. 8.8e). By reducing the thickness of the sample to within the penetration depth of the Kirkendall Effect, we eliminate the supporting bulk structure, resulting in a freestanding tri-layered NP Ag | Ag | Ag, illustrated by the rightmost image in Fig. 8.8e, without the loss of sacrificial materials. This synthesis method can be potentially extended to a family of reactions, where the anion in solution reacts directly with the metal electrode, forming an insoluble salt, as outlined in Eq. (8.3). In the follow-up study, we will demonstrate the fabrication of nanoporous Au using this method.

$$A + X^- \rightleftharpoons AX + e^- \tag{8.3}$$

8.4 Conclusions

In Chapter 8, a versatile chlorine-neutral etching method was presented and discussed to create nanoporous metals without losing sacrificial material. This is a fundamental improvement over many current nanoporous metal fabrication methods, that require significant amounts of sacrificial materials. By utilizing a Cl-containing electrolyte, a robust electrochemical plating-stripping reaction efficiently converts a nonporous Ag substrate into a tri-layered nanoporous Ag | Ag | Ag, taking advantage of the Kirkendall effect. We have demonstrated a temperature dependence of our technique, showing that performing the reaction at higher temperatures leads to the formation of larger pores and ligaments. Furthermore, this method can be extended to a family of reactions, wherein the electrode metal can electrochemically react directly with the ions in the electrolyte, thereby redefining selective etching for fabricating nanoporous metals. In the subsequent Chapter 9, we will extend this method to trilayer nanoporous Ag | Au | Ag.

Appendix 8.A

Additional data on the synthesis of NP- Ag | Ag | Ag trilayer on a thin silver substrate at 0–3°C.

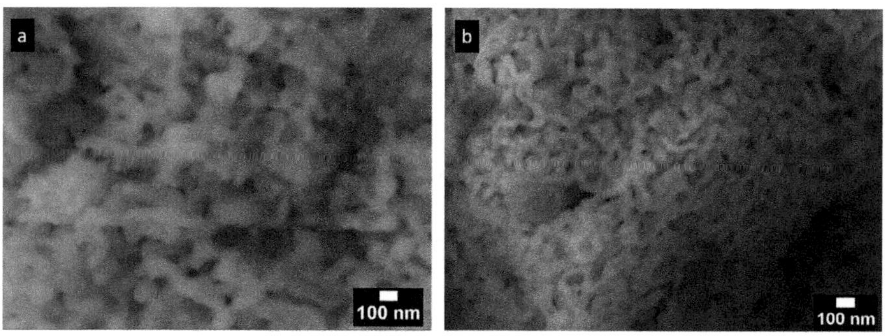

Figure A.8.1 Higher magnification SEM images of (a) the KE-NP-Ag (middle layer), synthesized at 0–3°C, and (b) the r-NP-Ag (outer layers), synthesized at 0–3°C. Average ligament sizes are measured to be ~180 nm and ≈80 nm, respectively.

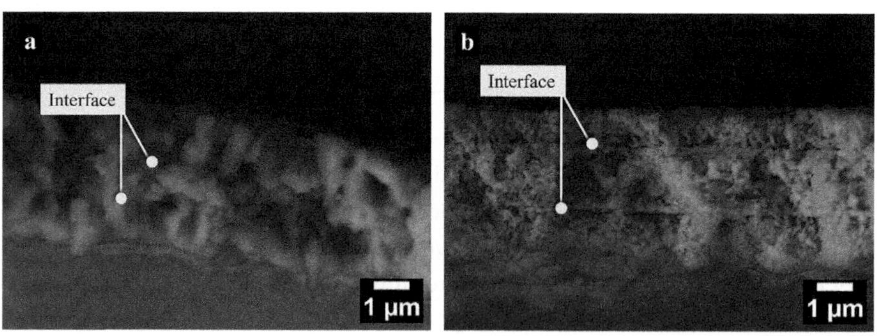

Figure A.8.2 Replicated images of cross-sectional images of the freestanding NP-Ag after (a) electrochemical plating, and (b) electrochemical stripping in an ice water bath (0°C). These are larger versions of the SEM images in Fig.4e and 4g. Notice the interfaces that appear, separating the KE-NP-Ag, and the outside layers of r-NP-Ag. Additional data on the synthesis of NP- Ag | Ag | Ag trilayer on a thin silver substrate at 25°C.

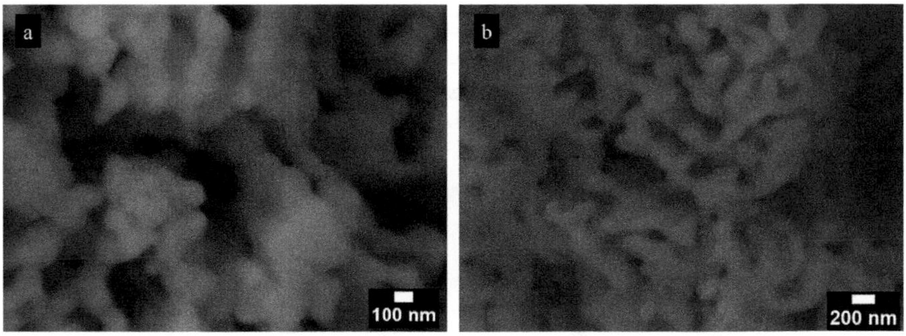

Figure A.8.3 Higher magnification SEM images of (a) the KE-NP-Ag (middle layer), formed during room-temperature synthesis, and (b) the r-NP-Ag (outer layers), formed during room-temperature synthesis. Average ligament sizes are measured to be$\sim$300 nm and $\approx$148 nm, respectively.

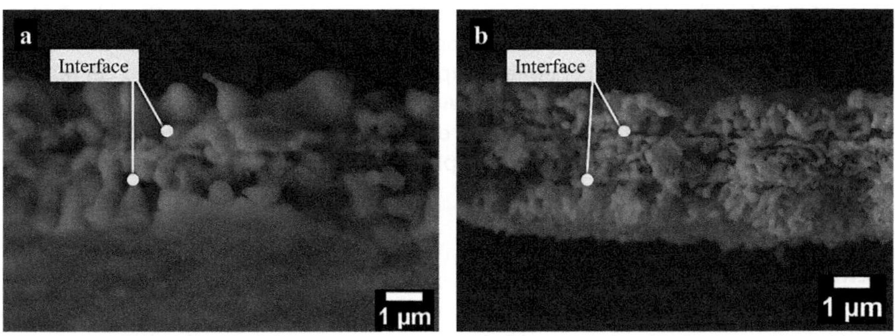

Figure A.8.4 Replicated images of cross-sectional images of the freestanding NP-Ag after (a) electrochemical plating, and (b) electrochemical stripping at room temperature (25°C). These are magnified versions of the SEM images in Figs. 8.4c and 8.5c. Notice the interfaces that appear, separating the KE-NP-Ag, and the outside layers of r-NP-Ag.

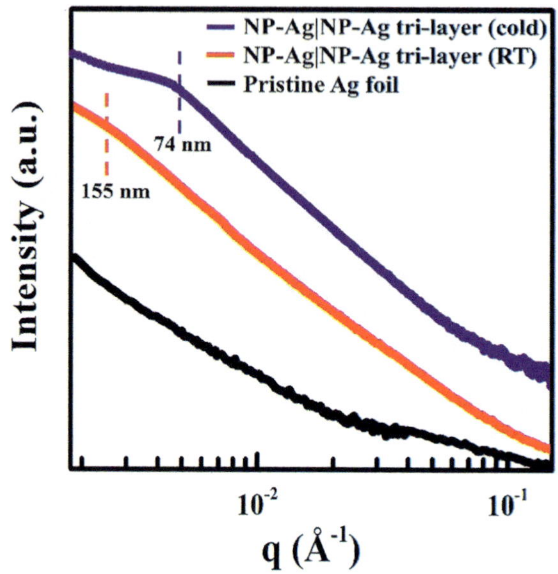

Figure A.8.5 Small-angle X-ray scattering (SAXS) curves comparing the thin film Ag leaf after *ice-water bath* electrochemical stripping of an AgCl layer to NP-Ag (blue curve) and after *room temperature* electrochemical stripping of the AgCl to NP-Ag (red curve). This can be compared to the nonporous dense Ag foil (black curve). Notice the beginning of a curve at the lower limit of the SAXS curve of the r-NP-Ag. In the main text, we describe how we used electrochemical plating/stripping at 0–3°C to shift the feature size within the SAXS measurement range.

Acknowledgments

The content of this chapter has been published in Acta Materialia 286 (2025) by Alexander K. Ng, Hyeongjun Koh, Eric Detsi, entitled, *Tri-layer nanoporous silver | silver | silver by etching without sacrificing materials through the Kirkendall effect*. The authors are thankful to Alexander K. Ng, Hyeongjun Koh for their discussion and contributions.

References

[1] Q. Chen, Y. Ding, M. Chen, Nanoporous metal by dealloying for electrochemical energy conversion and storage, MRS Bull. 43 (2018) 43–48, https://doi.org/10.1557/mrs.2017.300.

[2] X. Lang, A. Hirata, T. Fujita, M. Chen, Nanoporous metal/oxide hybrid electrodes for electrochemical supercapacitors, Nat. Nanotechnol. 6 (2011) 232–236, https://doi.org/10.1038/nnano.2011.13.

[3] Y. Yu, L. Gu, X. Lang, C. Zhu, T. Fujita, M. Chen, J. Maier, Li storage in 3D nanoporous Au-supported nanocrystalline tin, Adv. Mater. 23 (2011) 2443–2447, https://doi.org/10.1002/adma.201004331.

[4] Y. Gao, Y. Ding, Nanoporous metals for heterogeneous catalysis: following the success of Raney nickel, Chem. Eur J. 26 (2020) 8845–8856, https://doi.org/10.1002/chem.202000471.

[5] C.B. Rodella, G. Kellermann, M.S.P. Francisco, M.H. Jordão, D. Zanchet, Textural and structural analyses of industrial Raney nickel catalyst, Ind. Eng. Chem. Res. 47 (2008) 8612–8618, https://doi.org/10.1021/ie800543t.
[6] A.S. Hall, Y. Yoon, A. Wuttig, Y. Surendranath, Mesostructure-induced selectivity in CO_2 reduction catalysis, J. Am. Chem. Soc. 137 (2015) 14834–14837, https://doi.org/10.1021/jacs.5b08259.
[7] Q. Lu, J. Rosen, Y. Zhou, G.S. Hutchings, Y.C. Kimmel, J.G. Chen, F. Jiao, A selective and efficient electrocatalyst for carbon dioxide reduction, Nat. Commun. 5 (2014) 3242, https://doi.org/10.1038/ncomms4242.
[8] A.N. Koya, X. Zhu, N. Ohannesian, A.A. Yanik, A. Alabastri, R. Proietti Zaccaria, R. Krahne, W.-C. Shih, D. Garoli, Nanoporous metals: from plasmonic properties to applications in enhanced spectroscopy and photocatalysis, ACS Nano 15 (2021) 6038–6060, https://doi.org/10.1021/acsnano.0c10945.
[9] H.-J. Qiu, X. Li, H.-T. Xu, H.-J. Zhang, Y. Wang, Nanoporous metal as a platform for electrochemical and optical sensing, J. Mater. Chem. C 2 (2014) 9788–9799, https://doi.org/10.1039/C4TC01913J.
[10] E. Detsi, M. Salverda, P.R. Onck, J. Th, M. De Hosson, On the localized surface plasmon resonance modes in nanoporous gold films, J. Appl. Phys. 115 (2014) 044308, https://doi.org/10.1063/1.4862440.
[11] J. Weissmüller, R.N. Viswanath, D. Kramer, P. Zimmer, R. Würschum, H. Gleiter, Charge-induced reversible strain in a metal, Science 300 (2003) 312–315, https://doi.org/10.1126/science.1081024.
[12] H.-J. Jin, J. Weissmüller, D. Farkas, Mechanical response of nanoporous metals: a story of size, surface stress, and severed struts, MRS Bull. 43 (2018) 35–42, https://doi.org/10.1557/mrs.2017.302.
[13] H.-J. Jin, J. Weissmüller, A material with electrically tunable strength and flow stress, Science 332 (2011) 1179–1182, https://doi.org/10.1126/science.1202190.
[14] C. Zhu, G. Yang, H. Li, D. Du, Y. Lin, Electrochemical sensors and biosensors based on nanomaterials and nanostructures, Anal. Chem. 87 (2015) 230–249, https://doi.org/10.1021/ac5039863.
[15] E. Şeker, W.-C. Shih, K.J. Stine, Nanoporous metals by alloy corrosion: bioanalytical and biomedical applications, MRS Bull. 43 (2018) 49–56, https://doi.org/10.1557/mrs.2017.298.
[16] S. La Niece, Depletion gilding from third millennium BC Ur, Iraq 57 (1995) 41–47, https://doi.org/10.2307/4200400.
[17] M.F. Grimwade, The surface enrichment of carat gold alloys – depletion gilding, Gold Technol. 26 (1999) 16–23.
[18] J. Fu, S.S. Welborn, E. Detsi, Dealloyed Air- and water-sensitive nanoporous metals and metalloids for emerging energy applications, ACS Appl. Energy Mater. 5 (2022) 6516–6544, https://doi.org/10.1021/acsaem.2c00405.
[19] J. Fu, E. Detsi, J.Th.M. De Hosson, Recent advances in nanoporous materials for renewable energy resources conversion into fuels, Surf. Coat. Technol. 347 (2018) 320–336, https://doi.org/10.1016/j.surfcoat.2018.05.001.
[20] J. Fu, J.S. Corsi, Z. Wang, H. Wei, E. Detsi, Integrated metal-air battery and selective electrolytic leaching cell for the preparation of nanoporous metals, ACS Appl. Nano Mater. 1 (2018) 4164–4169, https://doi.org/10.1021/acsanm.8b00919.
[21] A. Wittstock, J. Biener, J. Erlebacher, M. Bäumer, Nanoporous gold from ancient technology to a high-tech material, Nanoporous Gold. https://doi.org/10.1039/9781849735285, 2012.
[22] X. Wang, Z. Qi, C. Zhao, W. Wang, Z. Zhang, Influence of alloy composition and dealloying solution on the formation and microstructure of monolithic nanoporous

silver through chemical dealloying of Al–Ag alloys, J. Phys. Chem. C 113 (2009) 13139–13150, https://doi.org/10.1021/jp902490u.
[23] J.S. Corsi, J. Fu, L. Wang, S.S. Welborn, Z. Wang, E. Detsi, Sacrificial silver recovery during nanoporous gold formation by electrolytic dealloying of gold-silver alloy, J. Electrochem. Soc. 169 (2022) 063501, https://doi.org/10.1149/1945-7111/ac6344.
[24] J. Fu, J.S. Corsi, S.S. Welborn, V. Basile, L. Wang, A.K. Ng, E. Detsi, Eco-friendly synthesis of nanoporous magnesium by air-free electrolytic dealloying with recovery of sacrificial elements for energy conversion and storage applications, ACS Sustain. Chem. Eng. 9 (2021) 2762–2769, https://doi.org/10.1021/acssuschemeng.0c08157.
[25] J.S. Corsi, J. Fu, Z. Wang, T. Lee, A.K. Ng, E. Detsi, Hierarchical bulk nanoporous Aluminum for on-site generation of hydrogen by hydrolysis in pure water and combustion of solid fuels, ACS Sustain. Chem. Eng. 7 (2019) 11194–11204, https://doi.org/10.1021/acssuschemeng.9b00481.
[26] C. Wang, Q. Chen, Elucidating the kinetic root of the evolution of the oriented nanoporous metal from reduction-induced decomposition, Chem. Mater. 33 (2021) 2604–2610, https://doi.org/10.1021/acs.chemmater.1c00316.
[27] C. Wang, Q. Chen, Reduction-induced decomposition: spontaneous formation of monolithic nanoporous metals of tunable structural hierarchy and porosity, Chem. Mater. 30 (2018) 3894–3900, https://doi.org/10.1021/acs.chemmater.8b01431.
[28] X. Jin, J. Lu, P. Liu, H. Tong, The electrochemical formation and reduction of a thick AgCl deposition layer on a silver substrate, J. Electroanal. Chem. 542 (2003) 85–96, https://doi.org/10.1016/S0022-0728(02)01474-2.
[29] A.D. Smigelskas, E.O. Kirkendall, Zinc diffusion in alpha brass, Trans. AIME 171 (1947) 130–142.
[30] D.A. Porter, K.E. Easterling, Phase Transformations in Metals and Alloys (Revised Reprint), third ed., CRC Press, 2009 https://doi.org/10.1201/9781439883570.
[31] X. Liu, X. Zhang, M. Kosmidou, M.J. Detisch, T.J. Balk, Synthesizing nanoporous stainless steel films via vacuum thermal dealloying, Metals 13 (2023) 1255, https://doi.org/10.3390/met13071255.
[32] Y. Yin, R.M. Rioux, C.K. Erdonmez, S. Hughes, G.A. Somorjai, A.P. Alivisatos, Formation of hollow nanocrystals through the nanoscale Kirkendall effect, Science 304 (2004) 711–714, https://doi.org/10.1126/science.1096566.
[33] J. Guo, A. Chan, W. Li, D.L. Fan, Kirkendall effect in creating three-dimensional metal catalysts for hierarchically porous ultrathin graphite with unique properties, Chem. Mater. 29 (2017) 4991–4998, https://doi.org/10.1021/acs.chemmater.7b01518.
[34] Q. Li, R.M. Penner, Photoconductive cadmium sulfide hemicylindrical shell nanowire ensembles, Nano Lett. 5 (2005) 1720–1725, https://doi.org/10.1021/nl050994x.
[35] M. Tanwar, H.K. Bezabh, S. Basu, W.-N. Su, B.-J. Hwang, Investigation of sodium plating and stripping on a bare current collector with different electrolytes and cycling protocols, ACS Appl. Mater. Interfaces 11 (2019) 39746–39756, https://doi.org/10.1021/acsami.9b10097.
[36] A.J. Crowe, J.L. DiMeglio, K.K. Stringham, B.M. Bartlett, Kinetics of magnesium deposition and stripping from non-aqueous electrolytes, J. Phys. Chem. C 121 (2017) 20613–20620, https://doi.org/10.1021/acs.jpcc.7b06068.
[37] I. Levin, NIST Inorganic Crystal Structure Database (ICSD). National Institute of Standards and Technology, https://doi.org/10.18434/M32147.
[38] K. Sieradzki, R.R. Corderman, K. Shukla, R.C. Newman, Computer simulations of corrosion: selective dissolution of binary alloys, Philos. Mag. A 59 (1989) 713–746, https://doi.org/10.1080/01418618908209817.

[39] I. McCue, B. Gaskey, P.-A. Geslin, A. Karma, J. Erlebacher, Kinetics and morphological evolution of liquid metal dealloying, Acta Mater. 115 (2016) 10–23, https://doi.org/10.1016/j.actamat.2016.05.032.

[40] Y.K. Chen-Wiegart, S. Wang, Y.S. Chu, W. Liu, I. McNulty, P.W. Voorhees, D.C. Dunand, Structural evolution of nanoporous gold during thermal coarsening, Acta Mater. 60 (2012) 4972–4981, https://doi.org/10.1016/j.actamat.2012.05.012.

[41] A.Y. Chen, S.S. Shi, F. Liu, Y. Wang, X. Li, J.F. Gu, X.F. Xie, Effect of annealing atmosphere on the thermal coarsening of nanoporous gold films, Appl. Surf. Sci. 355 (2015) 133–138, https://doi.org/10.1016/j.apsusc.2015.07.065.

[42] S.-H. Joo, J.W. Bae, W.-Y. Park, Y. Shimada, T. Wada, H.S. Kim, A. Takeuchi, T.J. Konno, H. Kato, I. V Okulov, Beating thermal coarsening in nanoporous materials via high-entropy design, Adv. Mater. 32 (2020), https://doi.org/10.1002/adma.201906160.

[43] L.H. Qian, M.W. Chen, Ultrafine nanoporous gold by low-temperature dealloying and kinetics of nanopore formation, Appl. Phys. Lett. 91 (2007) 083105, https://doi.org/10.1063/1.2773757.

[44] E. Detsi, Z. Vuković, S. Punzhin, P.M. Bronsveld, P.R. Onck, J.T.M.D. Hosson, Fine-tuning the feature size of nanoporous silver, CrystEngComm 14 (2012) 5402–5406, https://doi.org/10.1039/c2ce25313e.

[45] S.S. Welborn, E. Detsi, Small-angle X-ray scattering of nanoporous materials, Nanosc. Horiz 5 (2020) 12–24, https://doi.org/10.1039/C9NH00347A.

[46] Z. Qi, U. Vainio, A. Kornowski, M. Ritter, H. Weller, H. Jin, J. Weissmüller, Porous gold with a nested-network architecture and ultrafine structure, Adv. Funct. Mater. 25 (2015) 2530–2536, https://doi.org/10.1002/adfm.201404544.

[47] M. Graf, B.-N.D. Ngô, J. Weissmüller, J. Markmann, X-ray studies of nanoporous gold: powder diffraction by large crystals with small holes, Phys. Rev. Mater. 1 (2017), https://doi.org/10.1103/PhysRevMaterials.1.076003.

[48] N.F. Berk, Scattering properties of a model bicontinuous structure with a well defined length scale, Phys. Rev. Lett. 58 (1987) 2718–2721, https://doi.org/10.1103/PhysRevLett.58.2718.

[49] S. Niauzorau, A. Sharstniou, V.K. Sampath, N. Kublik, H. Bandarenka, B. Azeredo, Electroless dealloying of thin-film nanocrystalline Au–Ag alloys: mechanisms of ligament nucleation and sources of its synthesis variability, ACS Appl. Mater. Interfaces 14 (2022) 17927–17939, https://doi.org/10.1021/acsami.1c24388.

CHAPTER NINE

Nanoporous tri-layer of dissimilar elements by etching without sacrificing materials through the Kirkendall effect

9.1 Overview

In Chapter 8, we used the Kirkendall effect to demonstrate a novel sustainable method for fabricating metal/metal nanocomposites of the same element, such as tri-layer nanoporous Ag | Ag | Ag, without losing sacrificial materials. Here, we extend this new approach to create metal/metal nanocomposites of dissimilar elements, such as tri-layer nanoporous Ag | Au | Ag. Simply applying an oxidative and reductive current to an $Au_{35}Ag_{65}$ at.% alloy in a chloride-ion-containing electrolyte enables the plating and stripping of AgCl, resulting in a tri-layer nanoporous Ag | Au | Ag.

This sustainable synthesis approach not only reduces the loss of material during dealloying but also opens new avenues for developing metal/metal nanocomposites made of dissimilar nanoporous metal layers stacked together, offering significant implications for applications in catalysis and beyond, in which the synergistic effect between the underlying nanoporous metal layers could enhance the properties of the nanocomposite.

9.2 Dissimilar nanoporous metal layers

Metal/metal nanocomposites have been used in a wide range of applications such as catalysis [1—3], surface-enhanced Raman spectroscopy [4,5], and plasmonic heat generation [6], in which the synergistic effect between the underlying components enhances their properties [7]. For example, bimetal nanocomposites made of Au/Ag core-shell nanoparticles exhibit enhanced catalytic activity due to the synergic effect between Au and Ag [8—11]. As an extension, bimetal layered nanocomposites consisting of dissimilar nanoporous metal layers with high specific surface area, such as nanoporous Au and nanoporous Ag layers stacked together, represent another

Metals as Clean Fuels
ISBN: 978-0-443-13537-8
https://doi.org/10.1016/B978-0-443-13537-8.00004-X

class of metal/metal nanocomposites that can benefit from the synergistic effect between the underlying Au and Ag layers. Currently, though, layered nanoporous metal/metal composites are scarcely studied despite their documented improvements in performance [7,12]. We believe their complicated and nontrivial synthesis methods heavily contribute to the lack of research on these materials. An earlier report of layered nanoporous metal/metal composites has been made through bottom-up deposition, with several layers of precursor metal alloys being individually deposited through techniques like magnetron sputtering. The resulting material consists of different metal alloy layers, which can be placed in a corrosive solvent to selectively remove the sacrificial material (i.e., dealloying). The remaining material consists of layers of distinct nanoporous metals [12]. In another method, a metal is deposited in a sacrificial polymer template. The template is etched away using an e-beam, leaving behind a single nanoporous metal layer. This single nanoporous metal layer is used as the substrate for another polymer template for a different metal, resulting in a bi-layer of two dissimilar nanoporous metals [7]. These elaborate methods require specialized equipment. Therefore, methods to develop dissimilar layered nanostructured metals are desirable.

In Chapter 8, a sustainable *chlorine-neutral etching* method was presented without the loss of sacrificial materials, based on the Kirkendall effect, which we used to make tri-layer nanoporous silver | silver | silver (NP Ag | Ag | Ag) starting from a dense Ag foil [13]. In this work, we will demonstrate how this method can be extended to make tri-layer metal/metal nanocomposites with dissimilar nanoporous metals in each layer, such as tri-layer nanoporous silver | gold | silver (NP Ag | Au | Ag). We will use various microstructural characterization approaches, including electron microscopy and small-angle X-ray scattering, to study this tri-layer nanoporous system.

9.3 Experimental methods

9.3.1 Fabrication of freestanding tri-layer nanoporous Ag | Au | Ag film from a thin $Au_{35}Ag_{65}$ foil

To create a tri-layer nanoporous Ag | Au | Ag film (NP Ag | Au | Ag), we used a thin gold-silver parent alloy film with a thickness within the penetration depth of the Kirkendall effect. Previously, we found that during dealloying, the porosity extends 12 μm below the surface of the substrate [13]. Therefore, for this work, a $Au_{35}Ag_{65}$ at. % leaf (Sepp Leaf Products, New

York), with a thickness of approximately 6 μm is used, as it is below the thickness threshold. The electrochemical setup used here has been slightly modified from that in our last work, with the 6 μm $Au_{35}Ag_{65}$ leaf as the working electrode [13]. Aside from the working electrode, all other aspects of the cell configuration remained the same. To investigate the influence of temperature on the etching process, a series of experiments were conducted with the reaction vessel immersed in an ice-water bath (temperature measured at 0−3°C). This arrangement ensured that the electrolyte level within the reaction vessel remained lower than the level of the surrounding ice-water bath.

9.3.2 Materials characterization

X-ray diffraction (XRD) patterns were collected using a Rigaku Miniflex powder diffractometer (Cu X-ray source, operating at 40 kV/15 mA) with a 0.02 degrees step size and a 10 degrees/min scanning speed. A JEOL 7500F scanning electron microscope (SEM) equipped with an energy-dispersive X-ray spectroscopy (EDS) detector was used to characterize the microstructure of the etched $Au_{35}Ag_{65}$ at various stages of NP-Ag fabrication.

9.4 Results and discussion

9.4.1 Electrochemical plating-stripping model expanded to include Au

Below, we present a succinct model that illustrates the simplicity of the *chlorine-neutral etching* method. This method can be broken down into 3 steps: (1) Start with dense (i.e., nonporous): silver (Ag) foil in a dilute solution of sodium chloride (NaCl), illustrated in Fig. 9.1a (2) Apply a forward oxidative linear sweep voltammetry, or a positive current, to react Cl ions in solution with Ag, producing silver chloride (AgCl), through Eq. (9.1) (forward reaction). This process enlists the *Kirkendall effect* to extract Ag atoms from the bulk of the material which diffuse to the outer surface to react with Cl ion from the electrolyte, thus leaving behind vacancies in the bulk, which accumulate to form voids and, subsequently, a nanoporous network in the bulk [14]. As Ag atoms migrate to the substrate surfaces, they are converted to a dense layer of AgCl upon reaction with Cl ions from the electrolyte, forming an AgCl onto the Ag surface. This is illustrated as blue surface aggregates in the schematic in Fig. 9.1b and c (3) Apply a backward reductive linear sweep voltammetry, or a negative current, to

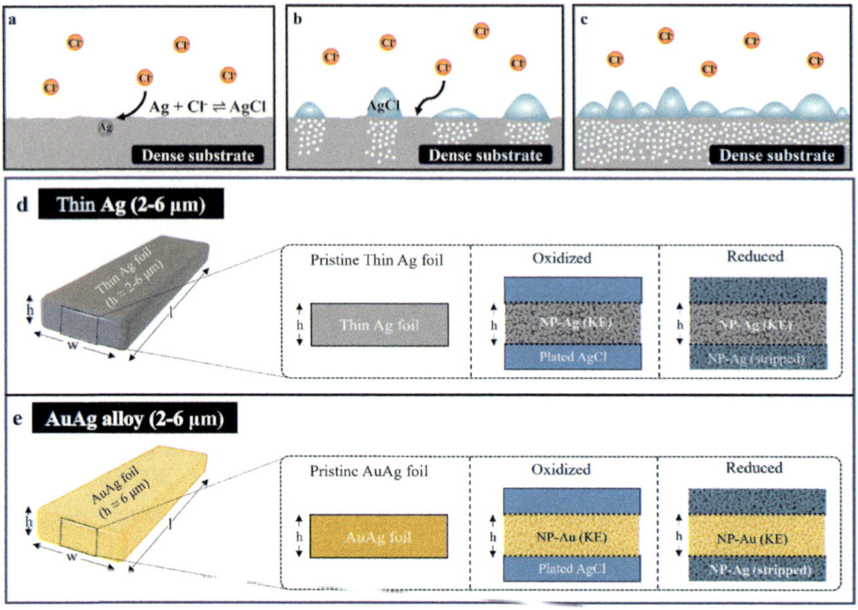

Figure 9.1 Step-by-step illustration of the mechanism through which electrochemical plating-stripping occurs during the fabrication of tri-layer nanoporous metals in three steps: (a) Ag atoms from the metal substrate react with Cl ions in the electrolyte. (b) AgCl nucleates at the reaction points and grows as the reaction proceeds along the surface. Ag atoms are also extracted from the bulk of the metal substrate through the Kirkendall effect. (c) AgCl aggregates continue to grow until the surface of the substrate is covered. (d) Formation of tri-layer nanoporous Ag | Ag | Ag; and I formation of tri-layer nanoporous Ag | Au | Ag from $Au_{35}Ag_{65}$ alloy.

reduce the AgCl formed at the substrate surface back to pure Ag through Eq. (9.1) (backward reaction), via *reduction-induced decomposition*, as illustrated in Fig. 9.1d. This method, developed by Wang et al., is defined by the fabrication of nanoporous metals through the chemical reduction of a nonmetallic compound [15].

In our work, we showed that we can reduce the AgCl we initially plated on the substrate back to NP-Ag. In Chapter 8, this technique was employed to develop both substrate-supported and freestanding nanoporous films. Because the Kirkendall effect is a diffusion-controlled process, the depth from which Ag atoms can diffuse and leave behind vacancies that cluster to form voids in the bulk Ag film is limited. Therefore, adjusting the thickness of the precursor Ag foil will control whether the resulting nanoporous silver (NP-Ag) will be a fully nanoporous film or substrate-supported.

Further details on the fabrication and characterization of the substrate-supported and freestanding NP-Ag can be found in Chapter 8 [13].

$$Ag + Cl^{-} \rightarrow AgCl + e^{-} \tag{9.1}$$

As shown in Fig. 9.1d, when a thin Ag foil is used as the precursor material, such that the thickness of the foil is within the penetration depth of the Kirkendall effect, the resulting foil is fully nanoporous, and composed of two parts: (i) a layer of Kirkendall effect-induced NP-Ag (KE-NP-Ag) shown in grey in Fig. 9.1d (most right illustration), sandwiched between (ii) two layers of reduction-induced NP-Ag (r-NP-Ag) shown in blue in Fig. 9.1d (most right illustration), which is a result of the reduction-induced decomposition of the plated AgCl layers at the substrate surface. This has been covered extensively in our preceding work [13]. Here, we aim to show that our chlorine-neutral etching process can be generalized to the family of reactions wherein a halide ion in solution reacts directly with dense metal to form the metal chloride, generalized in Eq. (9.2). We hypothesized that we could generalize this further by replacing the dense metal (A) with a bimetallic alloy (AB), wherein one of the metals in the alloy reacts with the halide ions in the electrolyte, as in Eq. (9.3). With this generalized model, we anticipate that our method can be used to make both nanoporous metal/metal composites with the same elements as well as nanoporous metal/metal composites with dissimilar elements, and tri-layer nanoporous metals.

$$A + X^{-} \rightleftharpoons AX + e^{-} \tag{9.2}$$

$$AB + X^{-} \rightleftharpoons (AX)B + e^{-} \tag{9.3}$$

Fig. 9.1e illustrates the proposed mechanism to explain the results of changing the precursor material from a dense metal (A) to a bimetallic alloy (AB)—we switched from using a dense Ag foil to a dense bimetallic Au_xAg_{1-x} at. % alloy as the precursor (where $x \lesssim 40\%$ represents the ratio between Au and Ag. Here $x = 35$). The $Au_{35}Ag_{65}$ foil is a homogeneous single-phase alloy of 6 μm thickness. Like in the pure Ag system, shown in Fig. 9.1d, we expect AgCl to form at the surface during oxidation, leaving a nanoporous Au center. This would be followed by reduction of the AgCl during reduction, converting the AgCl layer at the surface to nanoporous Ag. The following sections will be dedicated to characterizing and understanding this new structure.

9.4.2 Electrochemical plating-stripping on a dense $Au_{35}Ag_{65}$ alloy precursor

Because the chlorine-neutral etching process was originally performed using a dense Ag metal to create NP-Ag, we paired it with a comparable metal with low reactivity in air and water, namely a silver-rich gold-silver alloy with composition $Au_{35}Ag_{65}$ at. % to extend the concept to bimetallic systems. Furthermore, NP-Au and NP-Ag are two of the most well-characterized nanoporous metal systems [16—19]. Fig. 9.2 shows two CV curves—one in blue and one in red. The CV curve in blue was obtained when a dense Ag metal foil was used as the working electrode to make NP Ag | Ag | Ag tri-layer [13]. The CV curve in red is obtained when a dense $Au_{35}Ag_{65}$ alloy foil is used as the working electrode to make a tri-layer NP Ag | Au | Ag composite. The CV for each sample in Fig. 9.2 contains the oxidative forward sweep (forming AgCl from Ag and Cl ions) and the reductive reverse sweep (converting AgCl back to Ag), which were performed through a single CV at the scan rate of 0.1 mV/second between 0.35 and 0.8 V versus Ag/AgCl reference electrode. The conversion factor from Ag/AgCl reference electrode, stored in 0.1 M Cl ion solution, to SHE, is 0.29 V versus SHE.

Contrasting the two CVs, the red CV, with the $Au_{35}Ag_{65}$ precursor alloy as the working electrode, appears to have a shifted equilibrium potential (E_0)

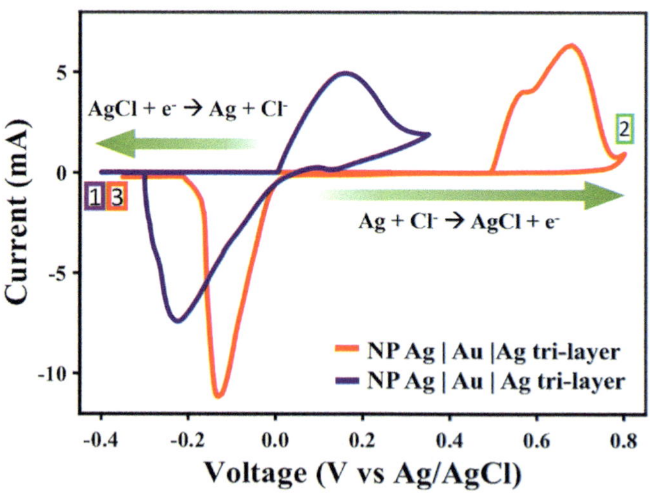

Figure 9.2 Cyclic voltammogram (CV) obtained during electrochemical plating-stripping on two different substrates—thin Ag substrate (blue) and thin $Au_{35}Ag_{65}$ alloy substrate (red).

from the original blue CV, with the pure Ag as the working electrode. The E_0 is typically estimated as the voltage that is halfway between the oxidative peak and reductive peak. To form the NP Ag | Ag | Ag system, blue CV, the E_0 is approximately −0.035 V versus Ag/AgCl. Conversely, to form the NP Ag | Au | Ag system, red CV, the E_0 is approximately 0.275 V versus Ag/AgCl. To ensure that this shift in E_0 is not caused by an unexpected side reaction at the electrode surface during electrochemical plating and stripping on the $Au_{35}Ag_{65}$ working electrode, we performed X-ray diffraction (XRD) to determine the crystal structure of the electrode at different states of charge during cycling.

X-ray diffraction (XRD) was used to analyze these samples at the following three oxidation/reduction states in the CV scan, highlighted in Fig. 9.2: pristine foil (state 1, purple box), after plating AgCl (state 2, green box), and after reduction of AgCl (state 3, red box). XRD results are compiled in Fig. 9.3. The XRD pattern of the pristine foil shows peaks that are between the expected values of Ag and Au (purple; ICSD Collection Codes #22434 and #52249, respectively) [20]. After plating with AgCl, peaks matching that of AgCl (green; ICSD Collection Code #56538) are introduced along with peaks of the substrate [20]. After reducing AgCl, the XRD peaks return to a mixture of Au and Ag (blue). Because Au and Ag are face-centered cubic

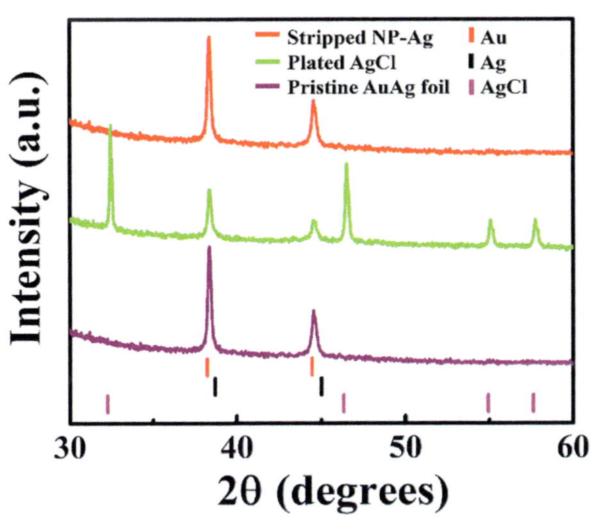

Figure 9.3 X-ray diffraction (XRD) partners collected on pristine $Au_{35}Ag_{65}$ substrate (purple), after plating of AgCl on the substrate (green), and after the AgCl layer back to Ag (red).

with only a small difference in lattice constants, the $Au_{35}Ag_{65}$ alloy and individual pure Au and Ag metals have nearly similar XRD patterns. The XRDs confirm that the expected products are formed at each step (AgCl formation after oxidation at step 2, and conversion to Ag after reduction at step 3), suggesting that the shift in E_0 that was shown in the red CV of Fig. 9.2 does not come from the formation of an unexpected phase. Considering this information, we can conclude that the difference in the CVs performed on the pure Ag electrode and the bimetallic $Au_{35}Ag_{65}$ alloy is mainly caused by an increase in the oxidation overpotential. This implies that switching the precursor material from a dense pure Ag foil to the $Au_{35}Ag_{65}$ alloy results in a change in the activation energy needed for the chlorine-neutral etching reaction to take place. In a simple bulk Ag system, Ag atoms can diffuse in any preferred direction as they move toward the substrate surface, where they are converted to AgCl. However, in the $Au_{35}Ag_{65}$ alloy, the Ag atoms are forced to diffuse around the existing Au atoms, which may form a highly tortuous diffusion pathway, thus hindering facile diffusion to the surface. As Ag atoms eventually reach the surface, they can react with the electrolyte to form AgCl. This phenomenon is realized during the synthesis method through the significant increase in overpotential that is required during the oxidation reaction, seen in the CV in Fig. 9.2.

9.4.3 Nanoporous silver | gold | silver tri-layer by electrochemical plating and stripping of AgCl on a thin $Au_{35}Ag_{65}$ substrate

We used scanning electron microscopy (SEM) and small-angle X-ray scattering (SAXS) to confirm the formation of the tri-layer nanoporous metal structure proposed in Section 9.1. We conducted the chlorine-neutral etching on a 6 μm-thick $Au_{35}Ag_{65}$ alloy as the Ag-rich substrate at room temperature (25°C, shown in this section) and in an ice-water bath (0–3°C, in the next section). Figs. 9.4 and 9.5 shows SEM images of the surface and cross section of samples after plating and reducing AgCl on the thin $Au_{35}Ag_{65}$ foil, at 25°C. SEM and EDS for the AgCl-coated sample's top surface (Fig. 9.4a and b) and the cross-section (Fig. 9.4c and d) are compiled in the orange box. SEM and EDS for the reduced sample's top surface (Fig. 9.5a and b) and the cross-section (Fig. 9.5c and d) are compiled in the blue box.

After plating AgCl on the surface of the $Au_{35}Ag_{65}$ leaf, there is a distinct interface between the AgCl layer and the NP-Au formed from the Kirkendall effect, (which we call KE-NP-Au), thus separating the $Au_{35}Ag_{65}$ leaf into three layers—one KE-NP-Au layer consisting of *ultrasmall pores* sandwiched

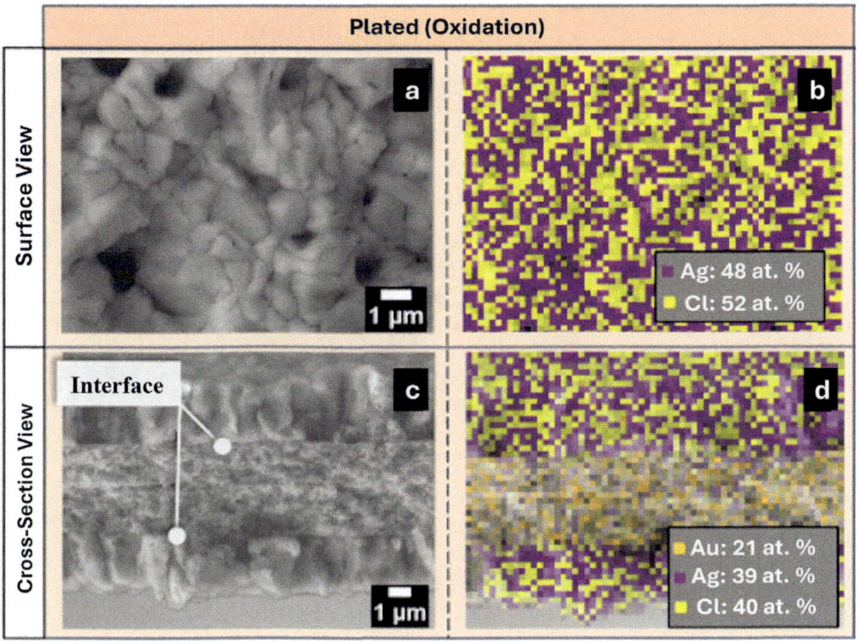

Figure 9.4 SEM, EDS, and SAXS of the $Au_{35}Ag_{65}$ alloy after electrochemical plating at room temperature. (a) Shows the surface view of the $Au_{35}Ag_{65}$ substrate after electrochemical plating. (c) Shows the cross-section of the $Au_{35}Ag_{65}$ after electrochemical plating. (b and d) Are the corresponding EDS maps that provide a map of the chemical composition of the given area. Ag is indicated by purple, Au is indicated by orange, and Cl is indicated by yellow.

between two layers of AgCl (Fig. 9.4c). The two sandwiching layers consist of dense AgCl aggregates, while the KE-NP-Ag middle layer has a nanoporous structure. SEM imaging normal to the surface, in Fig. 9.4a, reveals large aggregates of AgCl covering the entire substrate surface. The reduction step converts the AgCl back to Ag, resulting in layers of NP-Ag, formed by reduction-induced decomposition (which we call r-NP-Ag). The overall structure is the freestanding tri-layer NP-Au-Ag structure seen in Fig. 9.5c. Again, the interfaces demarcate three layers—one KE-NP-Au layer sandwiched between two layers of r-NP-Ag, as is confirmed by the EDS map in Fig. 9.5d. This tri-layer structure can be seen more clearly in Fig. A.9.2, where the cross-sectional SEM images from Figs. 9.4 and 9.5 are replicated large for easier visibility. According to the average measurements from the SEM images, the KE-NP-Au has an average ligament size of ≈ 9 nm (Fig. A.9.1b), while the r-NP-Ag has an average ligament size of ≈ 144 nm (Fig. A.9.1a).

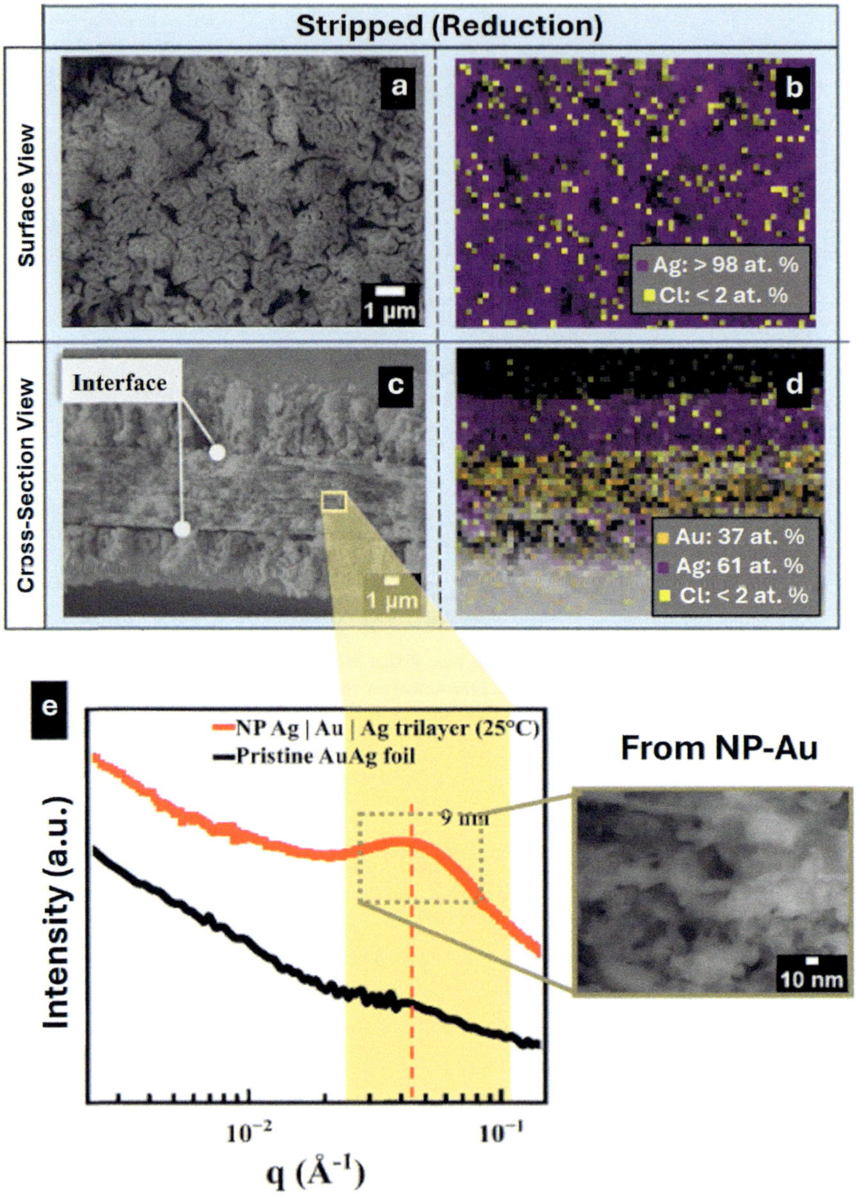

Figure 9.5 SEM, EDS, and SAXS of the $Au_{35}Ag_{65}$ alloy after electrochemical stripping at room temperature. (a) Shows the surface view of the $Au_{35}Ag_{65}$ substrate after electrochemical stripping. (c) Shows the cross-section of the $Au_{35}Ag_{65}$ after electrochemical stripping. (b and d) Are the corresponding EDS maps that provide a map of the chemical composition of the given area. Ag is indicated by purple, Au is indicated by orange, and Cl is indicated by yellow. (e) Compares the SAXS curves of the pristine $Au_{35}Ag_{65}$ precursor foil (black SAXS curve) and the resulting nanoporous metal tri-layer after the reduction step (red SAXS curve). The inset is a high magnification of the cross-section of the KE-NP-Au formed in the middle layer of the sample.

SAXS confirms that the porosity permeates throughout the nanoporous metal layers [21]. The peak position (q_{max}) in a SAXS curve relates to the ligament-ligament distance in a nanoporous metal through the approximation in Eq. (9.4) [22–24]. The ligament size is approximately half the ligament-ligament distance. We see a peak in the q-values at ≈ 0.045 Å^{-1}, which corresponds to approximately ≈ 9 nm. From the inset SEM image, we can conclude that this can only represent the KE-NP-Au formed by extracting Ag atoms from the bulk. The structure size of room-temperature dealloyed r-NP-Ag is outside the measurement range of our SAXS instrument. We expected to see SAXS peaks representing pores of ≈ 144 nm. As we see from Fig. 9.5e, this peak has been cut off by the lower measurement limits of our SAXS instrument. In Appendix 9.A, we present a plot (Fig. A.9.5) comparing the SAXS data from this electrochemical plating-stripping at room temperature and in an ice-water bath. In the next section, we demonstrate how changing the temperature will allow us to see the peak for the larger r-NP-Ag pores in the SAXS, as well.

$$q_{max} \approx 1.23 \frac{2\pi}{d} \tag{9.4}$$

To conclude this section, we highlight that the freestanding tri-layer composite NP Ag | Au | Ag structure is fabricated without special equipment or any pre- or post-processing steps. Using a commonly accessible NaCl electrolyte, we demonstrated that *sacrificial metals are no longer needed* to make NP-Au and NP-Ag. We can compare this to traditional nanoporous metals fabrication metals, where the more reactive metal is etched from a parent alloy, to leave behind a nanoporous structure of the more noble metal. To create a similar composite structure to that shown in Fig. 9.5c, would be a complicated process—involving individual layered depositions of Ag-containing and Au-containing parent alloys (such as $Al_{85}Ag_{15}$ and $Au_{35}Ag_{65}$), followed by subsequent dealloying steps. Each dealloying step removes greater than 60 at. % of the parent alloy to create the corresponding nanoporous structure (NP-Ag and NP-Au, respectively).

9.4.4 Effect of dealloying temperature on the morphology of the tri-layer nanoporous silver | gold | silver during electrochemical plating and stripping (0–3°C)

Thermal coarsening is widely known to impact the size of the pores in nanoporous metals [25,26]. Therefore, as mentioned in the previous section, we performed supplementary electrochemical plating-stripping experiments at

0–3°C, using an ice bath, to determine the effect of temperature on the feature sizes in our tri-layer nanoporous Ag | Au | Ag. Figs. 9.6 and 9.7 show the SEM and EDS data associated with the plating and reduction of AgCl on this thin $Au_{35}Ag_{65}$ foil at 0–3°C, respectively. SEM and EDS for the AgCl-coated sample top surface (Fig. 9.6a and b) and the cross-section (Fig. 9.6c and d) are compiled in the orange box. SEM and EDS for the reduced sample top surface (Fig. 9.7a and b) and cross section (Fig. 9.7c and d) are compiled in the blue box. Again, Ag atoms are extracted from the bulk of the alloy during the oxidation step, resulting in a middle layer of NP-Au with *ultrasmall pores* between AgCl aggregates (SEM image, Fig. 9.6c, confirmed in EDS, Fig. 9.6d). The reduction of the AgCl to Ag results in the freestanding tri-layer composite NP Ag | Au | Ag structure, confirmed by SEM and EDS (Fig. 9.7c and d respectively). Like in Figs. 9.4 and 9.5, these interfaces demarcate three layers—one NP-Au layer

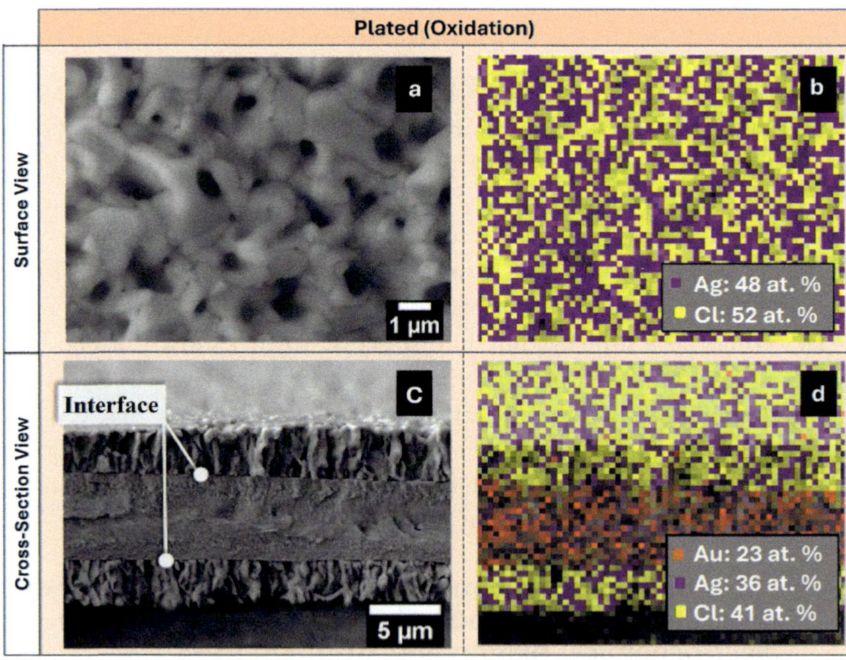

Figure 9.6 SEM and EDS images of the $Au_{35}Ag_{65}$ alloy after electrochemical plating and stripping in an ice-water bath. (a) Shows the surface view of the $Au_{35}Ag_{65}$ substrate after electrochemical plating. Notice the hierarchical structure. (c) Shows the cross section of the $Au_{35}Ag_{65}$ after plating. (b and d) Are corresponding EDS maps. Ag is indicated by purple, Au is indicated by orange, and Cl is indicated by yellow.

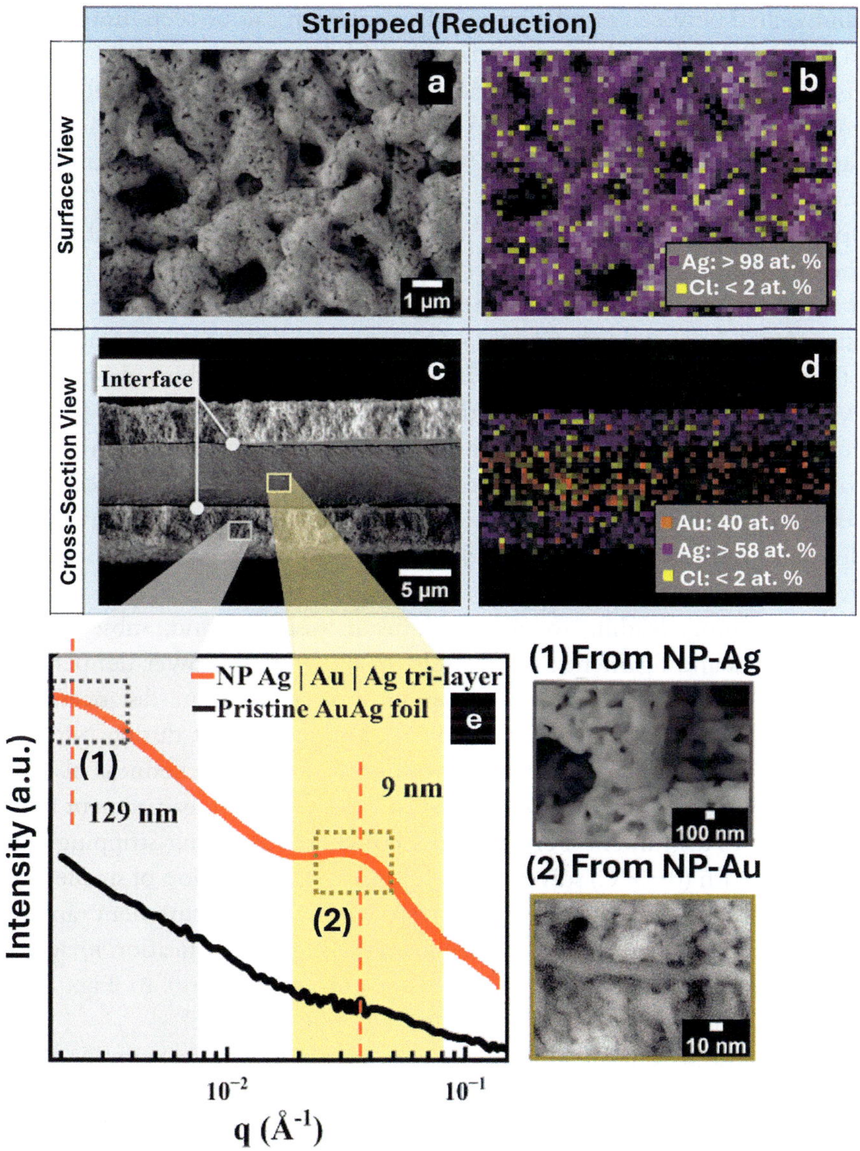

Figure 9.7 SEM and EDS images of the $Au_{35}Ag_{65}$ alloy after electrochemical plating and stripping in an ice-water bath. (a) Shows the surface view of the $Au_{35}Ag_{65}$ substrate after electrochemical stripping. Notice the hierarchical structure. (c) Shows the cross-section of the $Au_{35}Ag_{65}$ after stripping. (b and d) Are corresponding EDS maps. Ag is indicated by purple, Au is indicated by orange, and Cl is indicated by yellow. (e) Compares SAXS curves of the pristine $Au_{35}Ag_{65}$ precursor foil (black SAXS curve) and the resulting nanoporous metal tri-layer after the reduction step (red SAXS curve). Notice there are two peaks in the red SAXS curve. The insets are high magnification images of the relevant cross section—KE-NP-Au on the right, and r-NP-Ag on the left.

sandwiched between two layers of NP-Ag, which can be seen more clearly in Fig. A.9.4, where the cross-sectional SEM images from Figs. 9.6 and 9.7 are replicated at higher magnification. However, there is a noticeable difference in the feature size. The NP-Au remains similarly sized, with an average ligament size of $\approx$10 nm (Fig. A.9.3b), while the surface NP-Ag displays a bimodal hierarchical nanoporous structure. The smaller of these bimodal pores was measured at $\approx$84 nm (Fig. A.9.3a)—significantly smaller than the $\approx$144 nm seen during room-temperature chlorine-neutral etching.

Contrary to the nanoporous structures observed in the surface NP-Ag after the reduction of AgCl at 25°C (previous section), the surface view of the r-NP-Ag at 0−3°C (Fig. 9.7a) reveals a hierarchical bimodal porosity. We attribute this to the slower diffusion kinetics during lower temperature etching, which allows time for the development of preferred percolation pathways. As the rate of diffusion of the dealloying front is suppressed under colder conditions, it gives time for preferred percolation pathways to develop, resulting in the bimodal hierarchical structure. Furthermore, because dealloying is driven by surface diffusion, applying heat leads to thermal coarsening, leading to greater diffusion distances and, subsequently, larger pores and ligaments. Conversely, dealloying at a lower temperature can slow surface diffusion, resulting in shorter diffusion distances and, thus, smaller pores and ligaments [19,27]. We found that during NP Ag | Au | Ag tri-layer fabrication, room-temperature electrochemical plating/stripping yields larger pore sizes that even exceeded the measurement range of our SAXS instrument. Performing electrochemical plating-stripping in an ice-water bath (0−3°C) solves this, resulting in the formation of smaller features in the NP-Ag structure, which were within the measurement range of our SAXS instrument, as we can see from the high magnification images in Appendix 9.A, comparing Fig. A.9.1 (room-temperature) to Fig. A.9.3 (0−3°C), and confirmed by SAXS (in Figs. 9.5e and 9.7e).

SAXS was used to determine the characteristic length scale in this nanoporous structure. Shown in red in the SAXS plot (Fig. 9.7e), we can see two distinct peaks—corresponding to ligament sizes of $\approx$10 nm, from the ultra-small NP-Au pores, and $\approx$129 nm, from the NP-Ag pores (smaller of the bimodal pair), consistent with the SEM characterization of the NP-Au and r-NP-Ag layers, with higher magnification images in the insets. The larger of the bimodal pair, measured to have a ligament size of $\approx$1 μm from the SEM images, was outside the measurement range of our SAXS instrument, so the peak is not shown in Figs. 9.6 and 9.7.

Aligning with our current discussion, the 0—3°C r-NP-Ag displays a SAXS peak, while structure size of the 25°C r-NP-Ag was beyond the range of our SAXS instrument, indicating smaller pores throughout the cold-dealloyed film. Conversely, the structure size of the KE-NP-Au produced at different temperatures is relatively unchanged. We attribute this to the kinetic limitations in surface diffusion of dealloying of nanoporous gold. Typical coarsening experiments on nanoporous gold require extended periods of time at temperatures of several hundred degrees Celsius to induce thermal coarsening [25,28,29]. The 25°C difference that we are exploring in this work provides—not unexpectedly—insignificant energy in the milli eV range—making it reasonable to expect that the change in microstructure due to the 25°C difference would be barely noticeable.

9.5 Conclusions

In this Chapter 9 we have demonstrated the versatility of a novel chlorine-neutral etching method for creating tri-layer nanoporous metal/metal composites using dissimilar metals. By replacing the precursor material with a bimetallic alloy, we efficiently converted a dense $Au_{35}Ag_{65}$ substrate into a tri-layer nanoporous Ag | Au | Ag composite utilizing a Cl-containing electrolyte through chlorine-neutral etching, which leverages the Kirkendall effect and reduction-induced decomposition.

This method can be extended to a family of reactions, wherein any bimetallic alloy, having one metal in the alloy that can electrochemically react directly with the ions in electrolyte can be dealloyed to produce a tri-layer material of distinct nanoporous metals. By adjusting the temperature, our findings revealed that the feature size in r-NP-Ag is highly temperature-sensitive, whereas the feature size in KE-NP-Au remains relatively unaffected by small temperature changes. Through this innovative technique, we have redefined selective etching for the fabrication of nanoporous metals and introduced a simple, sustainable method for creating tri-layer nanoporous metal/metal nanocomposites.

Appendix 9.A

Room temperature (25°C) synthesis of NP-Au-Ag tri-layer from an $Au_{35}Ag_{65}$ substrate

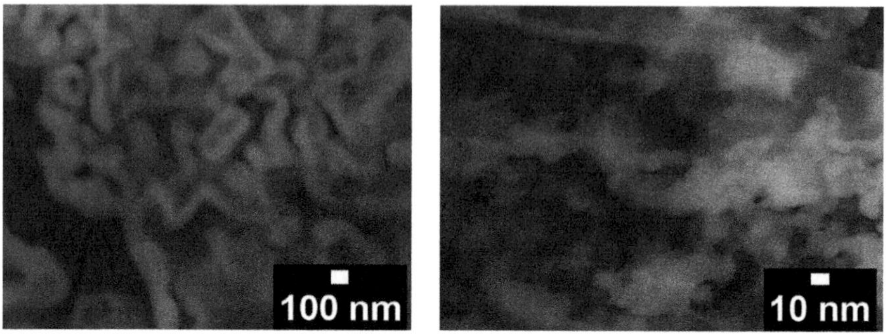

Figure A.9.1 Higher magnification SEM images of (a) the r-NP-Ag (outer layers), synthesized at room temperature (25°C), and (b) the KE-NP-Au (inner layer), synthesized at 25°C. Average ligament sizes are measured to be about 144 nm and about 9 nm, respectively.

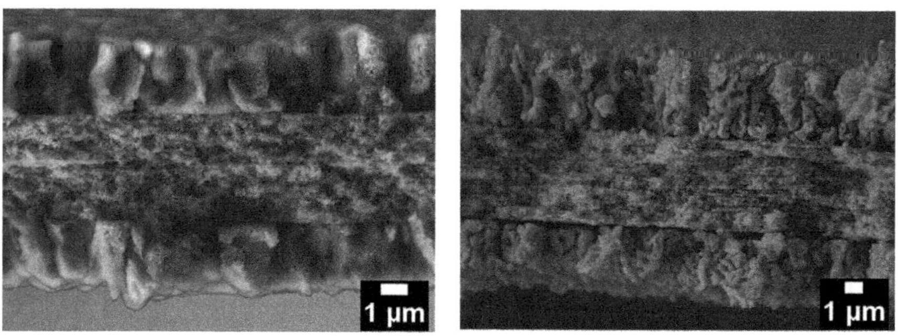

Figure A.9.2 Replicated images of cross-sectional images of the freestanding NP-Ag after (a) electrochemical plating, and (b) electrochemical stripping at room temperature. These are larger versions of the SEM images in Fig. 4e and 4g. Notice the interfaces that exist, separating the middle layer of KE-NP-Au, and the outside layers of r-NP-Ag.

Ice water bath (0°C) synthesis of NP-Au-Ag tri-layer from an $Au_{35}Ag_{65}$ substrate

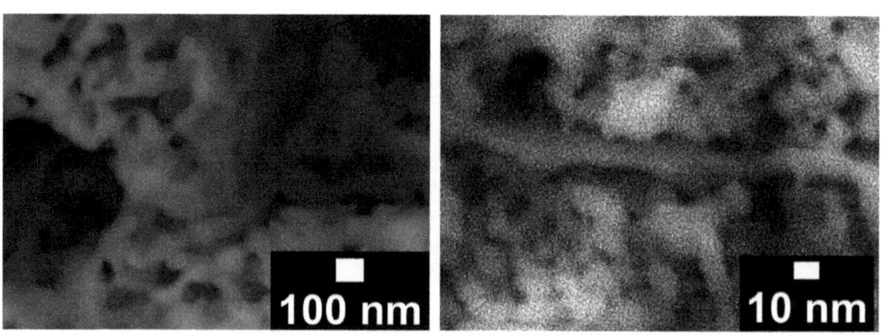

Figure A.9.3 Higher magnification SEM images of (a) the r-NP-Ag (outer layers), formed in an ice water bath (0°C) and (b) the KE-NP-Au (middle layer), formed in an ice-water bath. Average ligament sizes are measured to be about 84 nm and about 10 nm, respectively.

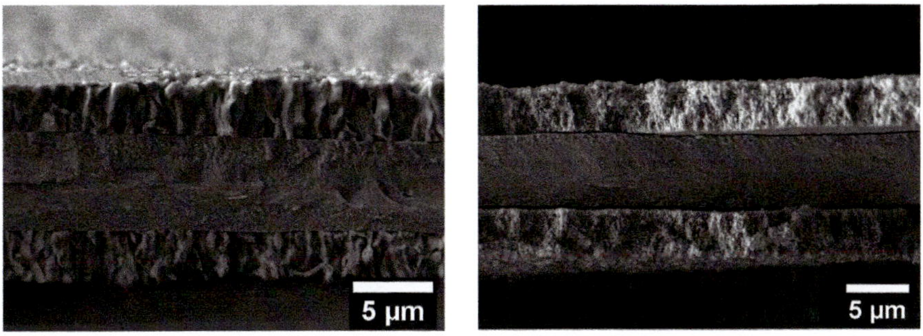

Figure A.9.4 Replicated images of cross-sectional images of the freestanding NP-Ag after (a) electrochemical plating and (b) electrochemical stripping at room temperature (25°C). These are magnified versions of the SEM images in Figs. 9.6c and 9.7c. Notice the interfaces that exist, separating the KE-NP-Ag, and the outside layers of r-NP-Ag.

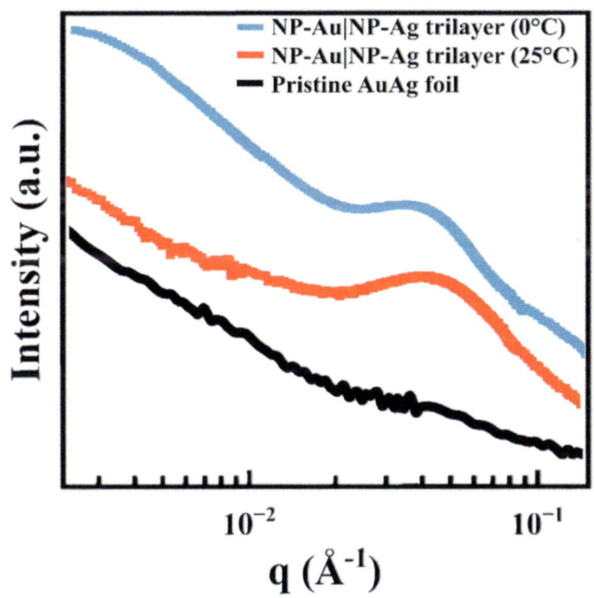

Figure A.9.5 Small-angle X-ray scattering (SAXS) curves comparing the thin film Ag leaf after ice-water bath electrochemical stripping of an AgCl layer to NP-Ag (blue curve) and after room temperature electrochemical stripping of the AgCl to NP-Ag (red curve). This can be compared to the nonporous dense Ag foil (black curve). Notice the beginning of a curve at the lower limit of the SAXS curve of the r-NP-Ag. In the main text, we describe how we used electrochemical plating/stripping at 0–3°C to shift the feature size within the SAXS measurement range.

Acknowledgments

The content of this chapter has been published in Acta Materialia 284 (2025) by Alexander K. Ng, Hyeongjun Koh, Eric Detsi, entitled *Tri-layer nanoporous silver | gold | silver by etching without sacrificing materials through the Kirkendall effect.* The authors are thankful to Alexander K. Ng, Hyeongjun Koh for their discussion and contributions.

References

[1] K. Sytwu, M. Vadai, J.A. Dionne, Bimetallic nanostructures: combining plasmonic and catalytic metals for photocatalysis, Adv. Phys. X 4 (2019), https://doi.org/10.1080/23746149.2019.1619480.

[2] M. Herran, S. Juergensen, M. Kessens, D. Hoeing, A. Köppen, A. Sousa-Castillo, W.J. Parak, H. Lange, S. Reich, F. Schulz, E. Cortés, Plasmonic bimetallic two-dimensional supercrystals for H_2 generation, Nat. Catal. 6 (2023) 1205–1214, https://doi.org/10.1038/s41929-023-01053-9.

[3] M. Herran, A. Sousa-Castillo, C. Fan, S. Lee, W. Xie, M. Döblinger, B. Auguié, E. Cortés, Tailoring plasmonic bimetallic nanocatalysts toward sunlight-driven H_2 production, Adv. Funct. Mater. 32 (2022), https://doi.org/10.1002/adfm.202203418.

[4] X. Hu, X. Wang, Z. Ge, L. Zhang, Y. Zhou, J. Li, L. Bu, H. Wu, P. Li, W. Xu, Bimetallic plasmonic Au@Ag nanocuboids for rapid and sensitive detection of phthalate plasticizers with label-free surface-enhanced Raman spectroscopy, Analyst 144 (2019) 3861–3869, https://doi.org/10.1039/C9AN00251K.

[5] Y. Su, S. Xu, J. Zhang, X. Chen, L.-P. Jiang, T. Zheng, J.-J. Zhu, Plasmon near-field coupling of bimetallic nanostars and a hierarchical bimetallic SERS "hot field": toward ultrasensitive simultaneous detection of multiple cardiorenal syndrome biomarkers, Anal. Chem. 91 (2019) 864–872, https://doi.org/10.1021/acs.analchem.8b03573.

[6] J. Gargiulo, M. Herran, I.L. Violi, A. Sousa-Castillo, L.P. Martinez, S. Ezendam, M. Barella, H. Giesler, R. Grzeschik, S. Schlücker, S.A. Maier, F.D. Stefani, E. Cortés, Impact of bimetallic interface design on heat generation in plasmonic Au/Pd nanostructures studied by single-particle thermometry, Nat. Commun. 14 (2023) 3813, https://doi.org/10.1038/s41467-023-38982-9.

[7] V. Caligiuri, H. Kwon, A. Griesi, Y.P. Ivanov, A. Schirato, A. Alabastri, M. Cuscunà, G. Balestra, A.D. Luca, T. Tapani, H. Lin, N. Maccaferri, R. Krahne, G. Divitini, P. Fischer, D. Garoli, Dry synthesis of bi-layer nanoporous metal films as plasmonic metamaterial, Nanophotonics 13 (2024) 1159–1167, https://doi.org/10.1515/nanoph-2023-0942.

[8] J. Huang, S. Vongehr, S. Tang, H. Lu, J. Shen, X. Meng, Ag dendrite-based Au/Ag bimetallic nanostructures with strongly enhanced catalytic activity, Langmuir 25 (2009) 11890–11896, https://doi.org/10.1021/la9015383.

[9] A.-Q. Wang, C.-M. Chang, C.-Y. Mou, Evolution of catalytic activity of Au–Ag bimetallic nanoparticles on mesoporous support for CO oxidation, J. Phys. Chem. B 109 (2005) 18860–18867, https://doi.org/10.1021/jp051530q.

[10] H. Wu, P. Wang, H. He, Y. Jin, Controlled synthesis of porous Ag/Au bimetallic hollow nanoshells with tunable plasmonic and catalytic properties, Nano Res. 5 (2012) 135–144, https://doi.org/10.1007/s12274-012-0194-5.

[11] Au-based bimetallic catalysts: how the synergy between two metals affects their catalytic activity, RSC Adv. 9 (2019), https://doi.org/10.1039/c9ra06001d.

[12] Y. Wang, S. Zhang, Z. Wu, Y. Fan, H. Chen, Q. Meng, Y. Yan, Z. Hu, J. Wang, E. Fu, Design and preparation of nanoporous Cu/Ag multilayer films, Coatings 11 (2021) 1187, https://doi.org/10.3390/coatings11101187.

[13] A. Ng, H. Koh, E. Detsi, Tri-layer nanoporous silver | silver | silver by etching without sacrificing materials through the Kirkendall effect, Acta Mater. 286 (2025) 120685.
[14] E. Kirkendall, A. Smigelskas, Zinc diffusion in alpha brass, Aime Trans. (1947) 130–142.
[15] C. Wang, Q. Chen, Reduction-induced decomposition: spontaneous formation of monolithic nanoporous metals of tunable structural hierarchy and porosity, Chem. Mater. 30 (2018) 3894–3900, https://doi.org/10.1021/acs.chemmater.8b01431.
[16] E. Seker, M.L. Reed, M.R. Begley, Nanoporous gold: fabrication, characterization, and applications, Materials 2 (2009) 2188–2215, https://doi.org/10.3390/ma2042188.
[17] M.D. Scanlon, U. Salaj-Kosla, S. Belochapkine, D. MacAodha, D. Leech, Y. Ding, E. Magner, Characterization of nanoporous gold electrodes for bioelectrochemical applications, Langmuir 28 (2012) 2251–2261, https://doi.org/10.1021/la202945s.
[18] Z. Zhang, Y. Wang, Z. Qi, C. Somsen, X. Wang, C. Zhao, Fabrication and characterization of nanoporous gold composites through chemical dealloying of two phase Al–Au alloys, J. Mater. Chem. 19 (2009) 6042–6050, https://doi.org/10.1039/B904052H.
[19] E. Detsi, Z. Vuković, S. Punzhin, P.M. Bronsveld, P.R. Onck, J.Th.M. De Hosson, Fine-tuning the feature size of nanoporous silver, CrystEngComm 14 (2012) 5402, https://doi.org/10.1039/c2ce25313e.
[20] I. Levin, NIST Inorganic Crystal Structure Database (ICSD), 2020, https://doi.org/10.18434/M32147.
[21] S.S. Welborn Ab, E. Detsi, Small-angle X-ray scattering of nanoporous materials, Nanoscale Horiz. 5 (2020) 12, https://doi.org/10.1039/c9nh00347a.
[22] Z. Qi, U. Vainio, A. Kornowski, M. Ritter, H. Weller, H. Jin, J. Weissmüller, Porous gold with a nested-network architecture and ultrafine structure, Adv. Funct. Mater. 25 (2015) 2530–2536, https://doi.org/10.1002/adfm.201404544.
[23] M. Graf, B.-N.D. Ngô, J. Weissmüller, J. Markmann, X-ray studies of nanoporous gold: powder diffraction by large crystals with small holes, Phys. Rev. Mater. 1 (2017), https://doi.org/10.1103/PhysRevMaterials.1.076003.
[24] N.F. Berk, Scattering properties of a model bicontinuous structure with a well defined length scale, Phys. Rev. Lett. 58 (1987) 2718–2721, https://doi.org/10.1103/PhysRevLett.58.2718.
[25] Y.K. Chen-Wiegart, S. Wang, Y.S. Chu, W. Liu, I. McNulty, P.W. Voorhees, D.C. Dunand, Structural evolution of nanoporous gold during thermal coarsening, Acta Mater. 60 (2012) 4972–4981, https://doi.org/10.1016/j.actamat.2012.05.012.
[26] A.Y. Chen, S.S. Shi, F. Liu, Y. Wang, X. Li, J.F. Gu, X.F. Xie, Effect of annealing atmosphere on the thermal coarsening of nanoporous gold films, Appl. Surf. Sci. 355 (2015) 133–138, https://doi.org/10.1016/j.apsusc.2015.07.065.
[27] L.H. Qian, M.W. Chen, Ultrafine nanoporous gold by low-temperature dealloying and kinetics of nanopore formation, Appl. Phys. Lett. 91 (2007) 2–5, https://doi.org/10.1063/1.2773757.
[28] S. Kuwano-Nakatani, T. Fujita, K. Uchisawa, D. Umetsu, Y. Kase, Y. Kowata, K. Chiba, T. Tokunaga, S. Arai, Y. Yamamoto, N. Tanaka, M. Chen, Environment-sensitive thermal coarsening of nanoporous gold, Mater. Trans. 56 (2015) 468–472, https://doi.org/10.2320/matertrans.MF201403.
[29] J. Wang, R. Xia, J. Zhu, Y. Ding, X. Zhang, Y. Chen, Effect of thermal coarsening on the thermal conductivity of nanoporous gold, J. Mater. Sci. 47 (2012) 5013–5018, https://doi.org/10.1007/s10853-012-6377-3.

CHAPTER TEN

Porous structures and their geometric and topological characteristics

10.1 Overview

Characterizing a porous materials system morphologically involves using both topological and geometrical descriptors to analyze the connectivity, content, and shape of porous patterns. This chapter presents and examines the fundamental principles of morphological image analysis based on integral geometry. It demonstrates that integral geometry offers these descriptors. This approach yields an objective, numerical characterization of two- and three-dimensional patterns in terms of geometrical and topological descriptors called Minkowski functionals.

In practice, the method of image analysis is based on functionals in *pixel* representations realizing that *each pixel* in a 2D(3D) image is a *convex set*. It is discussed that any morphological measure is a linear combination of the d+1 Minkowski functionals in *d*-dimensions (additive, motion invariant, continuous) and it works for any geometrical object. An illustrative example is given of applications of this approach to minimal surfaces.

As a more advanced application, we show how the technique can be used to obtain a morphological characterization of X-ray and electron microscopy (EM) tomography images of porous structures. It turns out that an extremely useful quantity to describe the porous system is the Euler characteristic χ, which is motion invariant and identical to the one defined in algebraic topology, where χ is directly related to the number of cavities and torus-like holes.

It is shown that integral-geometry morphological image analysis (MIA) can be applied directly to the digitized representation of patterns, can be executed with just a few lines of code, is computationally efficient, and is practical to use. It is demonstrated that MIA serves as an attractive toolset for analyzing geometrical objects and topological patterns in various porous systems.

Metals as Clean Fuels
ISBN: 978-0-443-13537-8
https://doi.org/10.1016/B978-0-443-13537-8.00005-1

10.2 Structural analysis

Structural analysis using X-rays and electron microscopy techniques in materials science primarily aims to connect microstructural observations with physical properties. The microstructural traits are influenced by chemical composition and processing methods, which means that comprehensive investigations of microstructures necessitate techniques that yield quantitative structural data. However, establishing a direct connection between structural analysis via X-rays and electron diffraction methods and the physical properties of a material is quite elusive. There are at least two reasons for this challenge. The defects that influence many physical properties often exist in a state that is not in thermodynamic equilibrium, and their behavior tends to be highly nonlinear. This poses a fundamental problem since there is a lack of robust physical and mathematical frameworks to thoroughly analyze these complex nonlinear and nonequilibrium phenomena. As a result, because various physical properties arise from the collective dynamic behavior of a defective solid, there is a critical need for quantitative three-dimensional information.

Another more practical reason why a quantitative evaluation of the structure-property relationship of materials is rather difficult has to do with statistics. Metrological considerations of quantitative electron diffraction from crystalline materials put some relevant questions to the statistical significance of the electron microscopy observations. Particularly, in situations where there is only a small volume fraction of defects present or a very inhomogeneous distribution, statistical sampling may be a problem. The usual approach to circumvent issues concerning statistics with electrons is to use 'volume-averaging' with X-rays in structural analysis. The common approach to describe and characterize spatial patterns is using the so-called structure factor $S(\vec{q})$, where $\vec{q}$ represents the X-ray scattering vector. Actually, the structure factor is an interference function, directly linked to the intensity I, and reads for a one-component isotropic system, Eq. (10.1):

$$I \propto \left\langle S\left(\vec{q}\right).S^{*}\left(\vec{q}\right)\right\rangle \propto \sum_{i,j}^{N} e^{-i\vec{q}.\left(\vec{r}_i-\vec{r}_j\right)} \tag{10.1}$$

By Fourier transforming $S(q) - 1$, a two-point correlation function $g(r)$ in real space over the atomic position vectors $\vec{r}_i$ in $3D$ is found, Eq. (10.2), i.e.,

$$g(r) - 1 \propto \int_0^{\infty} (S(q) - 1)\, q \sin qr\, dq \tag{10.2}$$

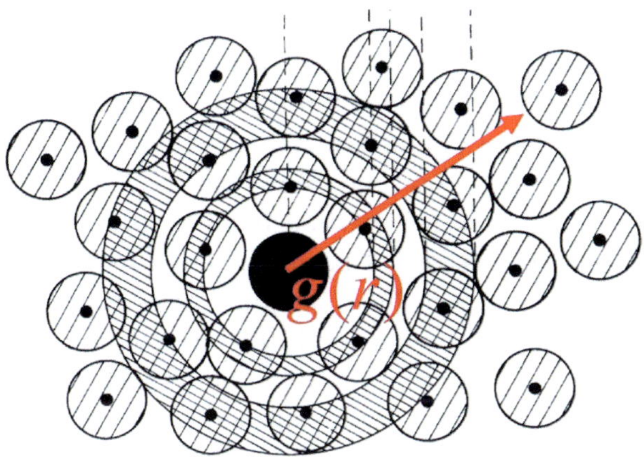

Figure 10.1 Radial distribution function obtained from X-ray structure analysis $g(r)$; Eq. (10.2) is a projection of 3D information onto; 1D radial distance of $\vec{r}$.

Obviously, the radial distribution function $g(r)$ from X-ray diffraction intensity does not provide information about the precise geometrical structure in 3D. By using the radial distribution function in the analysis, unfortunately, the 3D structural information causing the observed intensity I and $\left\langle S\left(\vec{q}\right).S^*\left(\vec{q}\right)\right\rangle$ is projected onto a 1D distance of $\vec{r}$ (see Fig. 10.1) and quantitative 3D structural information is completely lost. In fact, higher-order correlations are required. Indeed, those higher-order correlations, such as triplet angular correlations, can be revealed through X-ray analyses but only under very high external pressure, which is rather exotic experimentally.

In this chapter, we discuss another theoretical approach derived from integral-geometry-based morphological image analysis of highly porous structures where geometry and topology come together. It will be shown that the methodology yields a quantitative characterization of two- and three-dimensional patterns in terms of geometrical and topological descriptors called Minkowski functionals. The punchline is that in practice, image analysis can be based on functionals in *pixel* representations, realizing that *each pixel* in a 2D(3D) image is a *convex set*.

In general, the purpose of image analysis is to find out what is hidden in the data of these images [1–6] and to describe the data in quantitative physical numbers. In particular, we will analyze two-dimensional (2D) and three-dimensional (3D) patterns by numerical representations of the corresponding images in terms of two-valued Minkowski functionals and by making connections with geometry through Euler characteristics. Leonhard Euler observed that in any polyhedron, the sum of the number of vertices, v, and the number of faces, f, was two more than the number of edges [2,3]. Obviously, a numerical characterization requires that the image has been digitized, i.e., that the image has been converted into a numerical form [4–6]. This conversion may include additional digital image processing steps to enhance the quality of the images. If the image contains color or gray-level information, the digitization process should include the mapping of the spatial and color/brightness information in the image onto a collection of black-and-white image elements. For simplicity we will use the term *pixel* to refer to both 2D and 3D image elements.

In general, the study of morphology encompasses the form and structure of animals and plants in mathematical, physical, and biomedical sciences. It also refers to the examination of the geometry and topology of patterns. *Integral-geometry morphological image analysis* (MIA) utilizes additive image functionals to quantify the shape and connectivity of patterns created by the pixels in the image. Integral geometry [7–9] offers a precise mathematical framework for defining these image functionals. According to a fundamental theorem in integral geometry [7], the number of different additive image functionals is, under specific conditions, equal to the dimension of the pattern plus one. Thus, in the case of a 2D(3D) image, there are exactly 3 (4) of these functionals, called quermass integrals or Minkowski functionals. For a given image, the first step in MIA is to compute these functionals themselves. The second step is to study the behavior of the three or four numbers as a function of some control parameters, such as time, density, etc.

An outstanding characteristic of *integral-geometry morphological image analysis* is the significant gap between its ease of implementation plus usage and the advanced nature of the underlying mathematical theory. Indeed, as explained below, the calculation of the image functionals merely amounts to the proper counting of, e.g., faces, edges, and vertices of pixels. The application of MIA requires little computational effort.

Another appealing feature of MIA is that the image functionals have a geometrically and topologically clear interpretation: for 2D images, they correspond to the area, boundary length, and connectivity number. The four functionals for 3D images are the volume, surface area, integral mean curvature, and connectivity number.

This chapter gives an overview of the various aspects of MIA, with an emphasis on the practical application. MIA has proven to be very useful in describing the morphology in various fields of research, i.e., of porous media and complex fluids [10–13], the large-scale distribution of matter in the Universe [13–15], regional seismicity realizations [16], quantum motion in billiards [17], microemulsions [13,18], patterns in reaction diffusion systems [13,19], spinodal decomposition kinetics [13,20,21], and the dewetting structure in liquid crystal and liquid metal films [22], and in polymer films [23]. In many cases, additional understanding can be achieved by making assumptions about the dimensions, form, and arrangements of the objects. This generally involves developing a probabilistic model of the pattern and comparing its Minkowski functionals with those observed in the images. Applications of this stochastic-geometry approach to model natural phenomena can be found in Ref. [9].

In the previous introductory remarks so far, we assumed that the digitized images are free from any distortion, irregularities, and statistical noise that could impact the structure and layout of the objects of interest. Such flawless images can be easily created using computers and are highly beneficial for developing theoretical ideas and models. However, actual images or patterns obtained are rarely perfect. Thus, some type of image processing may be required before attempting to analyze the features in the image.

As said before, digital image processing is very relevant for many industrial, medical, and scientific applications. There is a vast amount of literature on this subject, so we can only cite a few books here [5,6]. There are also a huge number of different processing steps and methods. The type of measurements that will be performed on the image is an important factor in making a selection of the most appropriate processing steps.

The geometric and topological aspects of the image hold specific significance in integral-geometry morphological image analysis, and this should be evident in the methods used to improve image quality. Morphological image processing (MIP) technique [5,24–27] is suitable for this purpose as it shares mathematical foundations with MIA. In this chapter, we will concentrate solely on the issue of pattern characterization using morphological descriptors, without delving into the crucial aspect of image processing.

This chapter is structured as follows. Section 10.3 reviews the mathematical concepts of integral geometry as far as they are relevant to integral-geometry morphological image analysis (MIA). In Section 10.4, we describe how to use MIA in practice and provide algorithms to compute the morphological descriptors. In Section 10.5, we apply MIA to simple point patterns and geometrical objects (minimal surfaces) to illustrate the salient features of MIA. Section 10.6 discusses the application of MIA to computer tomography images of porous structures. Our conclusions and summary are given in Section 10.7.

10.3 Integral geometry: Theory

In this section, we examine the theoretical framework that forms the foundation of integral-geometry-based morphological image analysis. For simplicity, we adhere to the following conventions: We do not differentiate between two-dimensional and three-dimensional images; for instance, we will always use the term "pixel," even when referring to a "voxel." We focus solely on black-and-white images unless stated otherwise. The background color of an image is white, while the pixels that are filled in are black, making the others white. This is consistent with typical methods of creating and printing black-and-white images on paper. On a screen, of course, the roles of black and white are usually reversed.

10.3.1 Image measurements

The purpose of image analysis is to extract information from the image. Integral and differential geometry provides information on the geometrical and topological structure of sets of points in Euclidean space. In an image, the presence of a point in space is represented by a pixel that is black. Thus, objects in Euclidean space are represented by a collection of black pixels. Empty space corresponds to white pixels.

A function that assigns a number to a black-and-white image is called an *image functional* [26]. An image functional performs a measurement of certain properties or features in the image, such as the brightness, or location of objects, their surface, perimeter, size distribution, etc. An example of an image functional φ is the area of black pixels on a background of white pixels. If $\mathscr{P}_1$ and $\mathscr{P}_2$ are two patterns of black pixels, obviously we'll have

$$\varphi(\mathscr{P}_1 \cup \mathscr{P}_2) = \varphi(\mathscr{P}_1) + \varphi(\mathscr{P}_2) - \varphi(\mathscr{P}_1 \cap \mathscr{P}_2) \tag{10.3a}$$

where

$$\mathscr{P}_1 \cup \mathscr{P}_2 = \{p_1 \in \mathbb{R}^d :\ p_1 \in \mathscr{P}_1 \ .\ or\ .\ p_1 \in \mathscr{P}_2\} \tag{10.3b}$$

and

$$\mathscr{P}_1 \cap \mathscr{P}_2 = \{p_1 \in \mathbb{R}^d :\ p_1 \in \mathscr{P}_1 \ .\ \text{and}\ .\ p_1 \in \mathscr{P}_2\} \tag{10.3c}$$

In Eqs. (10.3), through $\mathscr{P}_1 \cup \mathscr{P}_2$ and $\mathscr{P}_1 \cap \mathscr{P}_2$, we refer to the *union* and the *intersection* of patterns $\mathscr{P}$, respectively, of sets $\mathscr{P}_1, \mathscr{P}_2 \subset \mathbb{R}^d$, in the d-dimensional Euclidean space $\mathbb{R}^d$ (*.and.* &*.or.* are so-called logical operators). As such $(\mathscr{P}_1 \cap \mathscr{P}_2)$, the last term in Eq. (10.3a), compensates for the double counting of black pixels that are common to $\mathscr{P}_1$ and $\mathscr{P}_2$.

Image functionals that share property Eq. (10.3) are called *additive*. Intuitively, it may seem obvious to require image functionals to be additive. In general, one would like to avoid counting a feature in an image more than once.

Additive image functionals are fundamental in integral geometry, but they are not the only beneficial image functionals. In fact, there are numerous image functionals that, while not additive, provide important insights into specific characteristics of an image [5,6,28]. A notable example is the two-point correlation function concerning the locations of black pixels (i.e., the Fourier transform of the structure factor, as shown in Eq. (10.2)). Though this is a nonadditive functional, it certainly offers valuable information about the spatial arrangement of the black pixels.

Another important property we would like to have is that the value of the image functional does not depend on the choice of the coordinate system. Formally, we say that an image functional φ is *motion invariant* if $\varphi(sA) = \varphi(A)$ for $s \in S$, where S denotes the group of all symmetry operations in the d-dimensional Euclidean space $\mathbb{R}^d$.

A key point of the theory reviewed in the subsections to come is that there are only $d + 1$ fundamentally different, motion-invariant, additive image functionals that describe the morphological content of a d-dimensional array of pixels. These functionals are directly related to simple geometrical concepts. The reader who is not interested in the mathematical aspects of integral geometry can skip the rest of this section and jump directly to Section 10.4.

10.3.2 Minkowski addition and subtraction

Consider two sets of points in Euclidean space (e.g., two images), specified by the position vectors $A = a_1, \ldots, a_n$ and $B = b_1, \ldots, b_m$. Translating every point of A by b_i, for $i = 1, \ldots, m$, and collecting all points yields a new set of points.

$$C = A \oplus B = B \oplus A \tag{10.4}$$

which is called the *Minkowski sum* of the sets A and B. The operation $\oplus$ is called *Minkowski addition*. If set B contains only one single point, say $b \in \mathbb{R}^d$ the Minkowski sum is a translation, $A \oplus \{b\} = A + b$.

Similarly, *Minkowski subtraction* $\ominus$ is defined by

$$C = A \ominus B = B \ominus A \tag{10.5}$$

That is, we translate every point of A by every element of B and only keep those points that intersect. Minkowski addition increases the number of points, while Minkowski subtraction removes points from an image. As such, the operations $\oplus$ and $\ominus$ play a central role in morphological image processing [26]. In this context, the former is often called *dilation* and has the effect of "inflating" the image A. The latter is called *erosion* and results in a "shrinking" of A.

It is noted that the Minkowski subtraction is not the "opposite," nor the reverse of the Minkowski addition, i.e., $A \ominus B \neq A \oplus (-B)$. For a review of the properties of Minkowski algebra, see, e.g., Refs. [5,7,26]. See also Appendix 10.A.

10.3.3 Parallel sets in Euclidean space

Consider the set of points of a line L of length a embedded in one-dimensional (1D) Euclidean space. We take a similar segment $S_r^{(1)}$ of length $2r$ and put the center of this line at each point of the line L. How does the union of all these points look like? Obviously, it is another line that is longer than L. The sets L (blue line), $S_r^{(1)}$ (orange line), and L_r (the result of this operation) are shown in Fig. 10.2.

In terms of the operations discussed above, it is clear that $L_r = L \oplus S_r^{(1)}$. The length l of L is given by

$$l(L_r) = a + 2r = l(L) + 2r. \tag{10.6}$$

The set L_r is called the parallel set of L at a distance r.

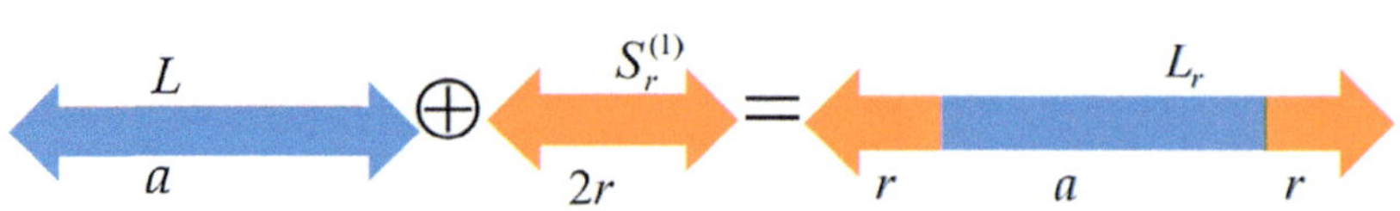

Figure 10.2 The Minkowski sum of a line segment L (depicted in blue) with a length of a and a line segment $S_r^{(1)}$ with a length of $2r$ produces the parallel set L (illustrated in blue and orange shading) at a distance of r.

The one-dimensional case readily extends to two and three dimensions. In Euclidean space, the *parallel set* A_r of A is defined by

$$A_r = A \oplus S_r^{(d)} \tag{10.7}$$

where $S_r^{(d)}$ denotes a d-dimensional sphere of radius r. Let us consider some simple examples, e.g., a circular disk D of radius a, a square Q of edge length a, and an equilateral triangle T of side length a embedded in the two-dimensional (2D) Euclidean space. Take a disc of radius r and perform the same operation as in the one-dimensional (1D) case. Put the center of the disc of radius r at each point of Q (of T or of D) and consider the union of all points. The resulting parallel sets Q_r, T_r, and D_r are shown in Fig. 10.3. The area U of Q_r, T_r and D_r is given by

$$U\left(Q_r\right) = a^2 + 4ar + \pi r^2. \tag{10.8a}$$

$$U\left(T_r\right) = \frac{\sqrt{3}}{4}a^2 + 3ar + \pi r^2. \tag{10.8b}$$

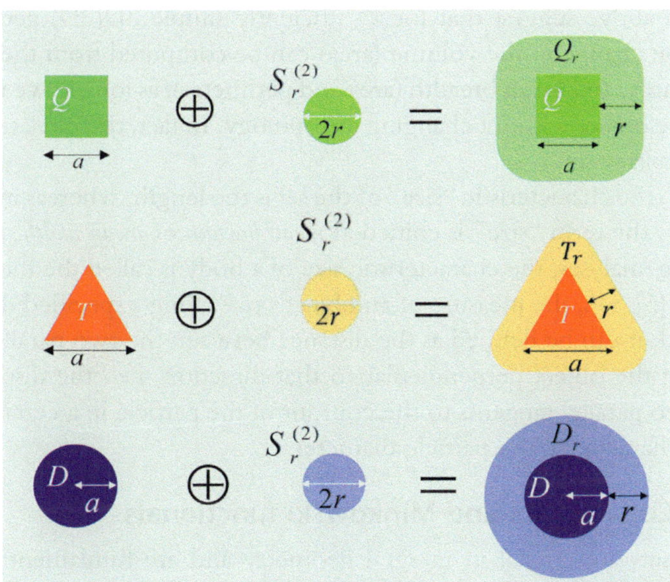

Figure 10.3 The Minkowski sum of a square Q of side length a, the sum of an equilateral triangle T of side length a, as well as the sum of a circular disk D with radius a, all three with a circular disc $S_r^{(2)}$ of radius r resulting in the parallel sets, Q_r, T_r, and D_r, respectively.

$$U\,(D_r) = \pi a^2 + 2\pi ar + \pi r^2. \tag{10.8c}$$

The Formulae (10.8) suggest that there may be a general relationship between the area of the original set and its parallel set at a distance *r*. It is not difficult to see that the areas of the three parallel sets can be written as

$$U(K_r) = U(K) + P(K)r + \pi r^2, \tag{10.9}$$

where *P* (*K*) denotes the boundary length (or perimeter) of the geometrical object *K*.

As a last example, we consider a cube *C* of edge length *a* embedded in three-dimensional (3D) space. A simple calculation shows that the volume *V* of the parallel set C_r can be written as

$$V(C_r) = a^3 + 6a^2r + 3a\pi r^2 + \frac{4\pi}{3}r^3. \tag{10.10}$$

Again, Eq. (10.10) suggests the generalization

$$V(K_r) = V(K) + S(K)r + 2\pi B(K)r^2 + \frac{4\pi}{3}r^3, \tag{10.11}$$

where *S*(*K*) is the *surface area* and *B*(*K*) is the *mean breadth*. The examples presented above suggest that for a sufficiently simple 3D(2D) geometrical object, the change in the volume (area) can be computed from the original volume, area, and mean breadth (area and perimeter), as long as we inflate or deflate the object without changing its topology. In fact, these relations only hold for *convex sets*.

In 1D, the characteristic "size" of the set is the length, whereas in integral geometry, the term "size" is coined as *mean breadth* or *mean width*. In image processing analyses, the characteristic size of a body is called the mean Feret diameter [5], i.e., the measure of an object's size along a specified direction. In general, it can be defined as the distance between the two parallel planes restricting the object perpendicular to that direction, i.e., the distance between two parallel tangents to the contour of the particle in a certain direction, to characterize the particle diameter.

10.3.4 Convex sets and Minkowski functionals

Convex sets are crucial in integral geometry and are fundamental to the morphological characterization of point sets in Euclidean space. It is clear that concepts related to these sets can also be applied to analyze collections of pixels. Nevertheless, for practical applications, there should be no

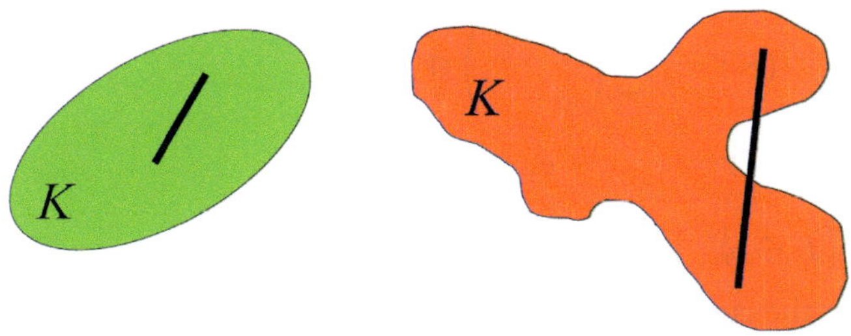

Figure 10.4 Definitions of convex and nonconvex bodies in *d*-dimensional Euclidean space $\mathbb{R}^d$; Left: Convex set, i.e., for every pair of points, the joining line also has to lie in *K*; Right: nonconvex, i.e., for not all pairs of points, the joining line lies fully in *K*.

limitations on the shapes of the objects involved. This section aims to explore the extension of the aforementioned concept to objects of arbitrary shapes. Formally, we define a set of points K in the d-dimensional Euclidean space $\mathbb{R}^d$ as a *convex set* if, for every two points in K, the entire line segment connecting them is also contained in K (see Fig. 10.4).

A convex set with a nonempty interior is called a *convex body*. A single point $\boldsymbol{x} \in \mathbb{R}^d$ is also a convex set and a convex body. We will only consider convex sets that are bounded and closed, i.e., that are *compact*. The class of all compact convex sets is denoted by $\mathbb{K}$.

The parallel set $K_r = K \oplus S_r^{(d)}$ of a compact convex set $K \in \mathbb{K}$ at a distance r is the union of all closed spheres of radius r, the centers of which are points of K [8]. The operation of taking a parallel set preserves the properties of convexity and compactness, i.e., $K_r \in \mathbb{K}$ [9].

The general expression for the volume V^d of the parallel body K_r at a distance r of a convex body K is given by the Steiner formula [7], where $W_\nu^{(d)}(K)$ are called *quermass integrals* or *Minkowski functionals*.

$$V^d(K_r) = \sum_{\nu=o}^{d} \binom{d}{\nu} W_\nu^{(d)}(K) r^\nu \quad r \geq 0, \tag{10.12a}$$

or

$$V^d(K_r) = \sum_{\nu=o}^{d} r^{d-\nu} \kappa_{d-\nu} V_\nu(K) \quad r \geq 0, \tag{10.12b}$$

or, more generally [5],

$$W_{\nu}^{(d)}(K) = \sum_{\mu=o}^{d-\nu} \binom{d-\nu}{\mu} W_{\mu+\nu}^{(d)}(K) r^{\mu} \quad r \geq 0, \qquad (10.12c)$$
$$\nu = 0, \ldots\ldots d,$$

where $V^d\left(K \oplus S_r^{(d)}\right)$ is the d-dimensional volume of K, and the parameter κ represents the curvature. The Minkowski functionals and intrinsic volumes are related to each other through Eqs. (10.12), and their relations to the familiar geometrical quantities are listed in Table 10.1. Clearly, Eq. (10.10) contains the results for the simple examples given in the previous subsection as special cases and is in accordance with Eqs. (10.12).

It can be shown [7] that the Minkowski functionals are motion invariant (i.e., do not depend on the choice of the coordinate system) and continuous, as well as convex C-additive, i.e., $\varphi(K_1 \cup K_2) = \varphi(K_1) + \varphi(K_2) - \varphi(K_1 \cap K_2)$ for $K_1, K_2 \in \mathbb{K}$ and $K_1 \cup K_2 \in \mathbb{K}$. The union of two convex sets is *not necessarily convex*, although *the intersection is*.

A fundamental result in integral geometry is the completeness of the family of Minkowski functionals. A theorem by Hadwiger [7] states that every motion invariant and continuous functional φ over $\mathbb{K}$ can be written as

$$\varphi(K) = \sum_{\nu=0}^{d} a_{\nu} W_{\nu}^{(d)}(K), \qquad (10.13)$$

with suitable coefficients $a_{\nu} \in \mathbb{R}$. A similar result holds if φ (K) is additive and monotonically increasing. Then the numbers a_{ν} are nonnegative. In other words, the $d + 1$ Minkowski functionals form a complete system of morphological measures on the class of convex sets $\mathbb{K}$ [7].

Table 10.1 Relationship between the Minkowski functionals $W_{\nu}^{d}(K)$ of a convex set K and conventional geometrical quantities.

	d = 1	d = 2	d = 3
$W_0^{(d)}(K)$	$\ell(K)$	$U(K)$	$V(K)$
$W_1^{(d)}(K)$	2	$P(K)/2$	$S(K)/3$
$W_2^{(d)}(K)$	–	π	$2\pi B(K)/3$
$W_3^{(d)}(K)$	–	–	$4\pi/3$

In summary, we may say that *integral-geometry morphological image analysis* (MIA) employs additive image functionals to represent images. Ideally, we prefer motion-invariant, additive image functionals.

If we could replace "C-additive" with "additive," Hadwiger's theorem would imply that there are precisely d + 1 unique additive image functionals. This would mean that we must explore nonadditive or coordinate system-dependent image functionals to discover further nonmorphological features within the image. Nevertheless, an image is not always a fully convex collection of points.

The broader application of Hadwiger's theorem to additive image functionals, instead of C-additive, requires further research. On a brighter note, it ultimately seems that notably Hadwiger's completeness theorem holds true for all categories of images as well.

10.3.5 Relation to topology and differential geometry

The Euler characteristic χ is identical to the one defined in algebraic topology [7]. For $d = 2$, $\chi(A)$ equals the number of connected components minus the number of holes. In three dimensions, $\chi(A)$ is given by the number of connected components minus the number of tunnels plus the number of cavities. Some examples are shown in Fig. 10.5. The Euler characteristic

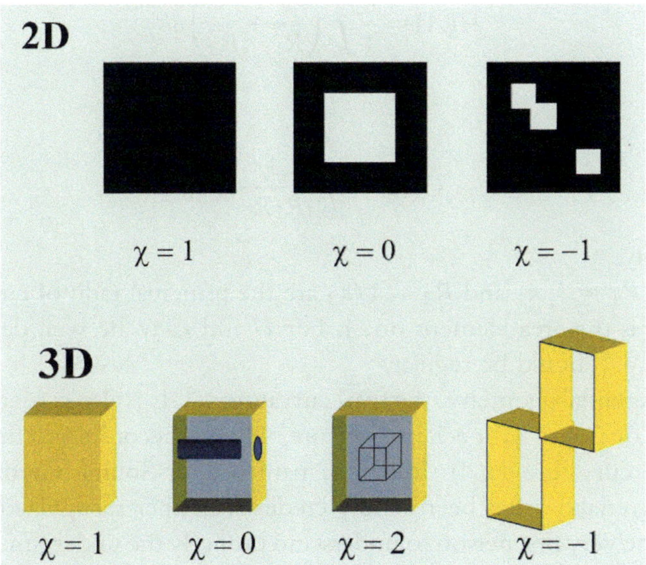

Figure 10.5 Two-dimensional and three-dimensional figures with various connectivity numbers (Euler characteristics) χ. *With permission of Ref. [4].*

describes A in a purely topological way, i.e., without reference to any kind of metric.

Very often, starting with Gauss and Euler, one is interested in the topology of the surface ∂A of A [2,14]. The Euler characteristic of ∂A is directly related to that of A, namely,

$$\chi(\partial A) = \chi(A)[1 - (-1)^n], \tag{10.14}$$

where n is the dimension of the body $A(n \leq d)$ [18].

The principal curvatures of a surface are important factors for the numerical analysis of a 3D body's surface. They are defined in the following way. Imagine a point on the surface along with the vector that is perpendicular to the surface at that point. A plane that contains this normal vector intersects the surface, creating a planar curve with a specific curvature known as the normal curvature. By rotating the plane around the normal, various planar curves emerge, each exhibiting different values of normal curvature.

The maximum and minimum values of these normal curvatures are referred to as the principal curvatures, i.e., κ_1 and κ_2 of a surface. These two curvatures can be combined to give two useful measures of the curvature of a surface, namely, the total Gaussian and mean curvature defined as $G = \kappa_1\kappa_2$ and $H = (\kappa_1 + \kappa_2)/2$, respectively. The integral mean curvature H and integral total curvature G are given by

$$H(A) = \frac{1}{2} \int_{\partial A} \left(\frac{1}{R_1} + \frac{1}{R_2} \right) df \tag{10.15}$$

and

$$G(A) = \int_{\partial A} \frac{1}{R_1 R_2} df, \tag{10.16}$$

respectively.

Here, $R_1 = 1/\kappa_1$ and $R_2 = 1/\kappa_2$ are the principal radii of curvature of A, and df is the area element on A. For H and G to be well defined, the boundary ∂A should be regular.

In differential geometry, the total curvature $G(A)$ is also coined Gaussian curvature or Gauss-Kronecker curvature, whereas according to the authors, the mean curvature $H(A)$ should be named after Sophie Germain [3,5], who—regrettably—has been hardly credited, neither recognized, for that. She was the very first person to understand properly the underpinning mathematics as well as the application of the mean curvature to open issues of elasticity in those days. As a follow-up to Euler for one-dimensional space, Sophie Germain suggested that in higher-dimensional space, the force of elasticity at a point is proportional to the sum of all the major curvatures at that point.

The narrative [3] reads that despite having established her "ingenious hypothesis," the initial equation she derived for a vibrating plane was incorrect. But the humorous shock is that her hypothesis itself directly leads to the accurate equation of a vibrating elastic plate $\ddot{w} = k^2\nabla^4 w$ as derived later by Lagrange from variational methods. Here, $\ddot{w}$ represents acceleration in a Newtonian force field description, i.e., the second derivative with respect to time of the out-of-plane deflection of a linear-elastic vibrational plate (x, y) . $\nabla^4. = \nabla^2\nabla^2$. is the biharmonic operator, and k is the ratio between the plate rigidity—depending on the modulus of elasticity and Poisson's ratio—and its mass (commonly written in terms of the density and thickness of the plate). From hindsight, it is rather remarkable that it took quite an effort by Siméon Poisson, Gustav Kirchhoff, and Claude-Louis Navier to get the correct value of k!

Sophie Germain's integral mean curvature $H(A)$ is proportional to the mean breadth B(A), see Section 10.3.3:

$$H(A) = 2\pi B(A). \tag{10.17}$$

The Euler characteristic of ∂A is closely related to the integral Gaussian curvature G and the genus g (number of handles, i.e., number of holes in the closed surface):

$$G(A) = 2\pi\chi(\partial A), \chi(\partial A) = 2(1-g). \tag{10.18}$$

Note that integral geometry imposes no regularity conditions on the boundary ∂A of the objects: $H(A)$ and χ (A) are always well-defined.

In addition to a description of the structure through the connectivity Euler number, likewise, Germain's curvature can be utilized to characterize the structure. In particular, the mean breadth, Eq. (10.17), can be used to get an approximate value of the specific ligament length ℓ in nano's or strut length s in macro foams, i.e., the mean total length per unit of volume, $L_v^{(\ell,s)}$. From Eqs. (10.12) and (10.17), and Table 10.1 follows:

$$\frac{2\pi}{3(d-1)} V_{d-2} = W_2^{(d)} = \frac{2}{3}\pi B(A) \tag{10.19}$$

or:

$$L_v^{(\ell,s)} \approx \frac{2B(A)}{\left(1 - V^{(l,s)}\right)} \tag{10.20}$$

The denominator reflects the porosity as derived from the density of the intrinsic volume. Eq. (10.20) has been used to correct for the loss of total strut/ligament length at the nodes [5].

10.3.6 Application to images

Every pixel in a 2D (or 3D) black-and-white image is a convex set. Thus, these images can be viewed as elements of the convex ring $\mathscr{R}$, allowing us to use integral geometry to create additive image functionals for measuring image features. However, as previously noted, it is essential to proceed with caution since Minkowski functionals are defined solely for convex sets. The foundation for effectively applying integral geometry to images lies in the property of additivity of χ: We can compute the Minkowski functionals of image A by decomposing A into convex sets K_i.

However, if we were to take for $\{K_i\}$ all black pixels (assuming the background consists of white pixels), then we would have to compute all the intersections. Although this can be done, it is much more expedient to take a slightly different route.

First, we write each pixel K as the union of the disjoint collection of its interior body, interior faces (in 3D only), open edges, and vertices [18]. We will denote the interior of a set A by $\widetilde{A} = A/\partial A$. The values of the Minkowski functionals of the open interior of an n-dimensional body $A \in \mathscr{R}$ embedded in $\mathbb{R}^d$ (n < d) are given by [18]

$$W_\nu^{(d)}\left(\widetilde{A}\right) = (-1)^{d+n+\nu} W_\nu^{(d)}(A); \ \nu = 0, \ldots, d. \tag{10.21}$$

By making use of the additivity of the Minkowski functionals and the fact that there is no overlap between open sets on a lattice, the values of the Minkowski functionals on the whole pattern $\mathscr{P}$ may be obtained from

$$W_\nu^{(d)}(\mathscr{P}) = \sum_m W_\nu^{(d)}\left(\widetilde{N}_m\right) n_m(\mathscr{P}) \ ; \ \nu = 0, \ldots, d, \tag{10.22}$$

where $n_m(\mathscr{P})$ denotes the number of open sets $\widetilde{N}_m$ of type m present in $\mathscr{P}$. On a square and cubic lattice, there are $d+1$ open sets $\widetilde{N}_m$: $\widetilde{N}_0$ corresponds to a vertex, $\widetilde{N}_1$ to an open line segment, $\widetilde{N}_2$ to an open square on both the 2D square and the 3D cubic lattice, and $\widetilde{N}_3$ to an open cube on the 3D cubic lattice.

Table 10.2 Minkowski functionals $W_\nu^{(2)}$ ($\nu = 0, \ldots, d = 2$) for the open sets $\widetilde{N}_m$, the basic building blocks of a square in two-dimensional Euclidean space: $\widetilde{Q}$ open square of edge length a; $\widetilde{L}$ open edge of length a; $\widetilde{P}$ vertex.

m	$\widetilde{N}_m$	$W_0^{(2)}$	$W_1^{(2)}$	$W_2^{(2)}$
0	$\widetilde{P}$	0	0	π
1	$\widetilde{L}$	0	a	$-\pi$
2	$\widetilde{Q}$	a^2	$-2a$	π

Table 10.3 Minkowski functionals $W_\nu^{(3)}$ ($\nu = 0, \ldots, d = 3$) for the open sets $\widetilde{N}_m$, the basic building blocks of a cube in three-dimensional Euclidean space.

m	$\widetilde{N}_m$	$W_0^{(3)}$	$W_1^{(3)}$	$W_2^{(3)}$	$W_3^{(3)}$
0	$\widetilde{P}$	0	0	0	$4\pi/3$
1	$\widetilde{L}$	0	0	$\pi a/3$	$-4\pi/3$
2	$\widetilde{Q}$	0	$2a^2/3$	$-2\pi a/3$	$4\pi/3$
3	$\widetilde{C}$	a^3	$-2a^2$	πa	$-4\pi/3$

$\widetilde{Q}$, open square of edge length a; $\widetilde{L}$, open edge of length a; $\widetilde{P}$, vertex; $\widetilde{C}$, open cube of edge length a.

The calculation of the Minkowski functionals for the building blocks $\widetilde{N}_m$ of a 2D square and a 3D cubic lattice is given elsewhere [4,29]. For convenience, their values are reproduced in Tables 10.2 and 10.3, respectively. The procedure to calculate $n_m(\mathscr{P})$ for a square and cubic lattice is given in Appendix 10.B to this chapter. Note that the value of the Minkowski functionals depends on the type of lattice, i.e., the shape of the pixels the pattern is represented on. A programming example (Fortran90) is listed in Appendix 10.C.

In Fig. 10.6, we show a simple example of the decomposition of the pixels of a two-dimensional black-and-white pattern (left) into squares, edges, and vertices (right). For this example, the number of squares $n_2 = 8$, the number of edges $n_1 = 24$, and the number of vertices $n_0 = 16$. Hence, $W_0^{(2)} = U = 8a^2$, $W_1^{(2)} = P/2 = -16a + 24a = 8a$, and $W_2^{(2)} = \pi\chi = 8\pi - 24\pi + 16\pi = 0$, where a denotes the edge length of the open square and open edge. Summarizing: In practice, the calculation of the Minkowski functionals of a black-and-white image amounts to counting vertices, edges, etc.

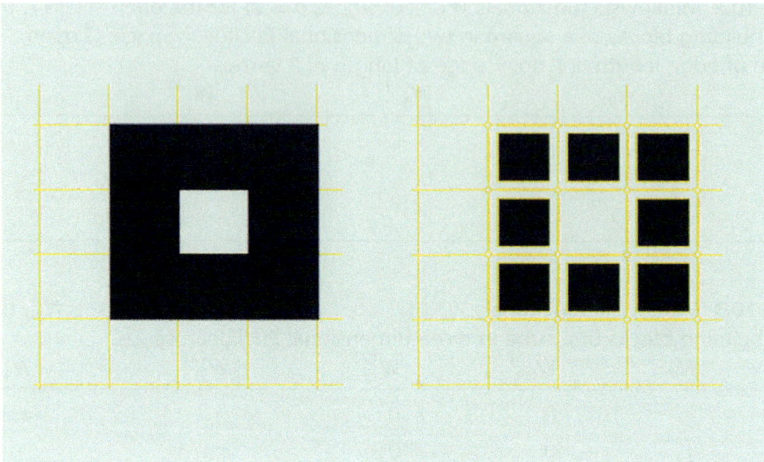

Figure 10.6 Illustration (left) breaks down the pixels of a two-dimensional black-and-white design into components such as squares, edges, and vertices (right).

10.4 Integral geometry in practice

One attractive aspect of the integral-geometry method is that the sophistication of the mathematical structure is not at all on par with the ease of its real-world usage. In this section, we emphasize the practical side of the approach. As a result, we will exclude the majority of mathematical validation and citations to the specific literature, but we will discuss a few examples. Our focus will solely be on black-and-white images.

From the theory reviewed above, it follows that the $d + 1$ Minkowski functionals form a complete system of additive image functionals on the set of objects that are unions of a finite number of convex sets [7]. In ordinary language, this implies that if we restrict ourselves to using additive image functionals, we only have to compute the $d + 1$ Minkowski functionals to characterize the morphology of the pattern.

In Appendix 10.C, we describe a straightforward and effective method for calculating the Minkowski functionals. Conceptually, more details are provided in Refs. [4,29]. In the following two subsections, we convert these concepts into a computer program. When transforming these mathematical ideas into a practical application, it is crucial to consider the conditions under which the aforementioned theorems are valid. Each pixel in a 2D(3D) black-and-white image is a convex set. Therefore, such images may be considered as an element of the convex ring $\mathscr{R}$ [7] and we can invoke integral geometry to build additive image functionals to measure features in the image. However, as mentioned before, some care has to be taken because the Minkowski functionals take known values on convex sets only. The crucial step is to decompose the image into a union of convex sets so that we can use the theoretical results that hold on the convex ring.

In the following subsections, we concentrate on the second phase of Integral Geometry Morphological Image Analysis (MIA): examining how these functionals rely on specific control parameters. The subsequent subsections address various technical subjects that are significant for practical applications.

10.4.1 Minkowski functionals

Imagine a two-dimensional grid populated with black squares against a white backdrop (refer to Fig. 10.6). To simplify, we assume that these squares have a uniform size of one unit each. Our goal is to analyze the geometry and topology of the arrangement created by the black squares. Based on Hadwiger's completeness theorem, there are three additive image functionals known as Minkowski functionals that describe the morphological content of this 2D pattern, namely the area U, the perimeter P, and the Euler characteristic χ. The latter describes the connectivity (topology) of the pattern: In 2D, χ equals the number of regions of connected black pixels minus the number of completely enclosed regions of white pixels. Two black pixels are considered "connected" if they are either nearest neighbors, or next-nearest neighbors, or they can be linked by a sequence of black pixels that are nearest or next-nearest neighbors. Based on this definition, we determine that the Euler characteristic of the pattern illustrated in Fig. 10.6 is zero. Conceptually, the procedure (that easily extends to three dimensions) to compute these three numbers consists of two steps. First, we decompose each black pixel into four vertices, four edges, and the interior of the pixel (see Fig. 10.6). Then we count the total number of squares n_2, edges n_1, and vertices n_0.

As aforementioned, an image functional, called (A), depends on pattern A described by an area of black pixels on a background of white pixels (see also Eqs. (10.3) and (10.13)). In its turn, any pattern A can be decomposed into objects, while any additive and motion-invariant functional φ must be a linear combination of the form:

$$\varphi(A) = \sum_{\nu=0}^{d} a_\nu F_\nu(A), \tag{10.23}$$

where F_ν plays the role of Minkowski functional. By definition, we have $F_\nu = n_\nu$ for $\nu = 0, \ldots, d$. Using Eq. (10.22) and Table 10.2 (2D), or 10.3 in 3D, it follows that the area U, perimeter P, and Euler characteristic χ are given by (see also Appendices 10.B and 10.C):

$$U = n_2, \quad P = -4n_2 + 2n_1, \quad \chi = n_2 - n_1 + n_0. \tag{10.24}$$

For the example displayed in Fig. 10.6, the number of squares $n_2 = 8$, the number of edges $n_1 = 24$, and the number of vertices $n_0 = 16$ and we find $U = 8$, $P = 16$ and $\chi = 0$.

For a 3D cubic lattice filled with black and white pixels, the four Minkowski functionals are the volume V, the surface area S, the mean breadth B (see Section 10.3), and the Euler characteristic χ. In 3D, χ equals the number of regions of connected black pixels plus the number of completely enclosed regions of white pixels minus the number of tunnels, i.e., regions of white pixels piercing regions of connected black pixels. As in the 2D case, the first step in the calculation of these four numbers is to consider each black pixel as the union of 8 vertices, 12 edges, 6 faces, and the interior of the cube. From Eq. (10.22) and Table 10.3, it follows that

$$\begin{aligned} V &= n_3; \\ S &= -6n_3 + 2n_2; \\ 2B &= 3n_3 - 2n_2 + n_1; \\ \chi &= -n_3 + n_2 - n_1 + n_0, \end{aligned} \tag{10.25}$$

where n_3 is the number of cubes. Thus, as in the 2D case, the morphological characterization of a 3D pattern reduces to the counting of the elementary geometrical objects (vertices, edges, faces, and cubes) that constitute the pattern.

10.4.2 Analysis of point patterns

Numerous systems can be represented by point patterns. For instance, a collection of particles can be interpreted as a set of points determined by their positions. These points are commonly referred to as the germs of the model [9,24]. To analyze the morphological characteristics of the point set (such as randomness, clustering, periodic arrangements, etc.), it is beneficial to associate discs (or spheres) of radius r to the points. These discs (or spheres) are known as the grains of the model [9,24]. Investigating how the image is covered by the grains provides insights into the distribution of the germs.

When we map the point pattern onto a square (cubic) lattice, we obtain a black-and-white image. The black pixels indicate the germs of the model. On this pixel lattice, we can create the grains of the model using two different approaches. In the first approach, we treat the germs as discs (spheres) with a radius $r = 0$. By increasing the radius to $r > 0$, we color all pixels that are within a distance of r from the germs in black.

The resulting grains provide discrete representations of discs (spheres) in Euclidean space. An example of this graining procedure in two dimensions is shown on the left-hand side of Fig. 10.7 for grains of radius $r = 3$. The right-hand side of Fig. 10.7 illustrates the second graining procedure (for $d = 2$), where we take the germs to be squares (cubes) of edge length $r = 1$ and the grains to be enlarged squares (cubes) of edge length $r = 2n + 1$, $n > 0$. Note that the growing of the cubic grains leads to a faster complete coverage of the image than the growing of the circular grains.

For this category of problems, MIA consists of the calculation of the three (four) numbers U, P, and χ (or V, S, B, and χ) as a function of the grain size r. A schematic representation of this procedure for the case of 2D point patterns is shown in Fig. 10.8.

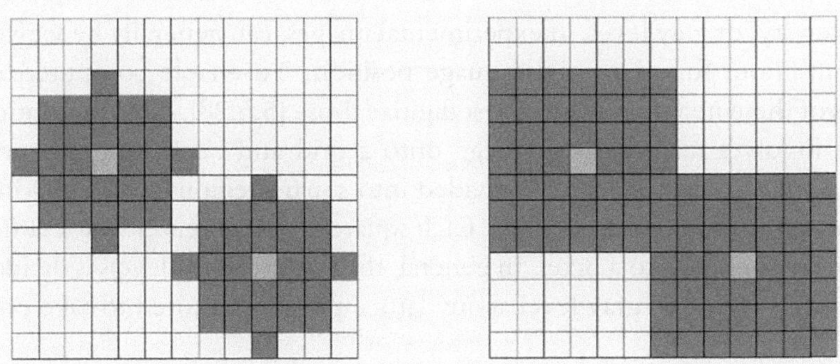

Figure 10.7 Graining procedure of a point pattern in two dimensions. Left: The grains are discrete approximations of a sphere with a radius of three in Euclidean space. Right: The grains are squares of edge length seven. The light gray pixels indicate the positions of the germs. *With permission of Ref. [4].*

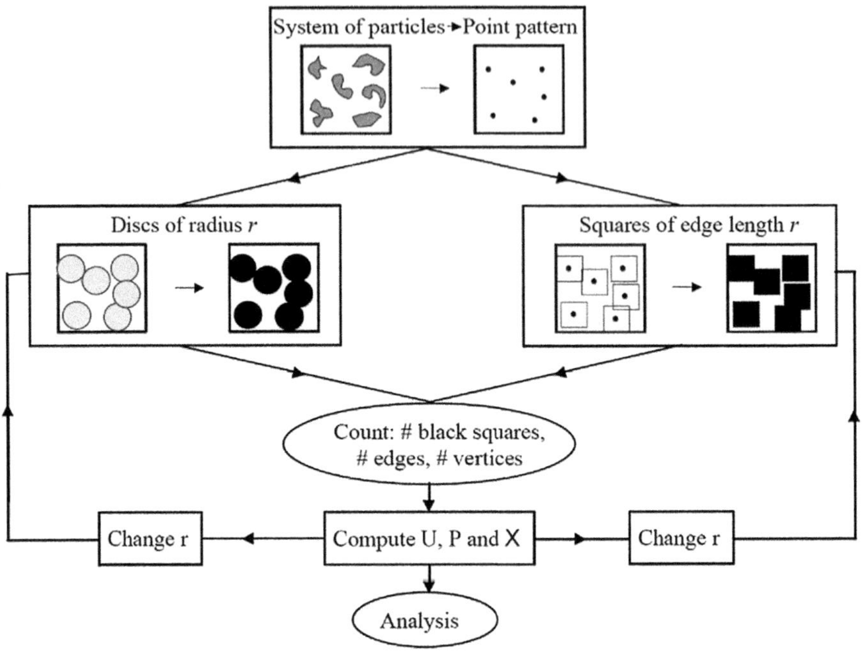

Figure 10.8 Schematic representation of how to use MIA to analyze the morphological properties of (patterns that can be interpreted in terms of) 2D point patterns. *With permission of Ref. [4].*

10.4.3 Analysis of digitized and threshold images

The intensity, or gray level, in experimental images, can generally be viewed as a continuous function of the image position. To enable computerized analysis of these images, we must first digitize those [5,6,28]. The digitization process involves mapping the image onto a grid and quantizing the gray levels. Typically, 2D images are divided into square regions, while 3D images are divided into cubic sections. Each square (cube) is centered at a lattice point, corresponding to a pixel. In general, the range of gray levels is divided into intervals, and the gray level at any lattice point is required to take only one of these values.

The output of image analysis should be a description of the given image. Thus, we have to define the various objects building up the image, i.e., we need a method to distinguish objects from the background [4–6].

The simplest method of reducing gray-scale images to two-valued images or black-and-white images is to make use of a *threshold*. If the given image $\mathscr{P}(\boldsymbol{x})$ with $\boldsymbol{x} \in \mathbb{R}^d$ has gray level range $[a, b]$, and q is any number between a and b, the result of thresholding $\mathscr{P}(\boldsymbol{x}, q)$ at q is the two-valued image $\mathscr{P}(\boldsymbol{x}, q)$ defined by Refs. [4–6]

$$\mathscr{P}(\boldsymbol{x}, q) = \begin{cases} 1\ \mathscr{P}(\boldsymbol{x}) \geq q \\ 0\ \mathscr{P}(\boldsymbol{x}) < q \end{cases} \tag{10.26}$$

By definition, if $\mathscr{P}(\boldsymbol{x}, q) = 0$, $\boldsymbol{x}$ is part of the background, and if $(\boldsymbol{x}, q) = 1$, $\boldsymbol{x}$ is part of an object. In practice, not all thresholds q yield useful $\mathscr{P}(\boldsymbol{x}, q)$. If q is too large, too many objects are classified as background or if q is too small, the opposite happens. Other thresholding operations may also be considered [4–6]. For this type of image, MIA consists of the calculation of the three (four) numbers U, P, and χ (or V, S, B, and χ) as a function of the threshold q. A schematic representation of this procedure for the case of 2D gray-scale images is shown in Fig. 10.9.

10.4.4 What integral geometry is incapable of doing

Hadwiger's theorem regarding the completeness of the d + 1 Minkowski functionals is quite significant, but the notion of "completeness" shouldn't

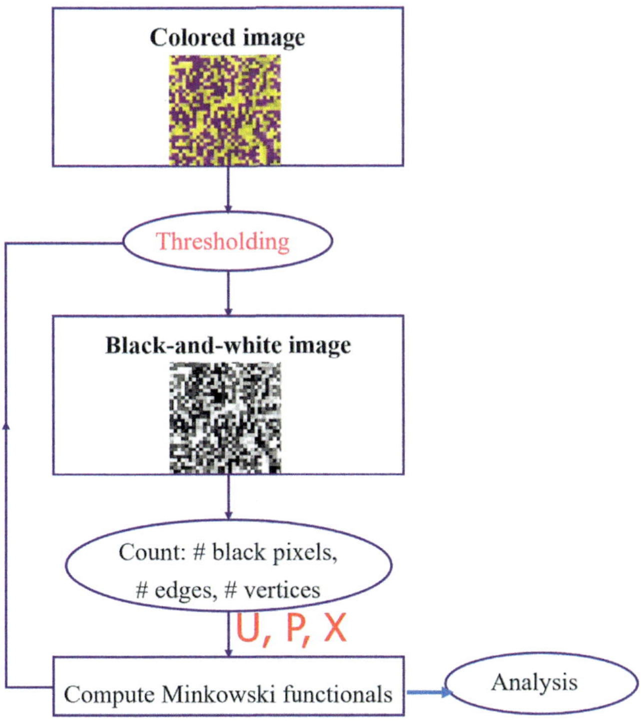

Figure 10.9 Schematic representation of how to use MIA to analyze the morphological properties of 2D digitized images.

be interpreted too strictly. A simple example can help clarify this concept. Consider a triangle situated in 2D Euclidean space. Can we use the Minkowski functionals to "completely" characterize this triangle? The answer is negative .

It is not difficult to show that the length of the edges of the triangle are given by

$$a = \frac{P}{4} + \frac{U}{P}\frac{1+\cos\theta}{\sin\theta} \pm \sqrt{\left(\frac{P}{4} + \frac{U}{P}\frac{1+\cos\theta}{\sin\theta}\right)^2 - \frac{2U}{\sin\theta}} \tag{10.27a}$$

$$b = \frac{P}{4} + \frac{U}{P}\frac{1+\cos\theta}{\sin\theta} \mp \sqrt{\left(\frac{P}{4} + \frac{U}{P}\frac{1+\cos\theta}{\sin\theta}\right)^2 - \frac{2U}{\sin\theta}} \tag{10.27b}$$

$$c = \frac{P}{2} - \frac{2U}{P}\frac{1+\cos\theta}{\sin\theta} \tag{10.27c}$$

where θ denotes the angle between the sides of lengths a and b, and U and P are the area and perimeter, respectively. From Eqs. (10.27), it is clear that the Minkowski functionals U, P, and $\chi = 1$ do not specify the triangle completely.

Different choices of the angle θ result in different triangles with the same area U and perimeter P. For *Integral-Geometry Morphological Image Analysis* (MIA) to yield useful information about the morphological content of an image, it is necessary to study the dependence of the Minkowski functionals on one or more control parameters (e.g., threshold, grain size, etc.).

10.4.5 Reducing digitization errors

Often, a grid of pixels represents an image of objects in Euclidean space. The Minkowski functionals calculated from this grid serve as approximations to those in Euclidean space. By digitizing a 2D (or 3D) image, we introduce square (or cubic) distortions in the objects, which create a directional bias. For instance, converting a 2D (or 3D) image changes a smooth contour (or surface) into a more segmented contour (or surface). The more intricate the image, the more effective the digital approximations tend to be, as the stepwise boundaries or surfaces will frequently display each orientation. Isotropic structures may pose the greatest challenges.

Various methods exist to tackle systematic inaccuracies that arise from the digitization of images [6,25]. Many of these techniques can improve the accuracy of estimations concerning area, perimeter, and other measurements.

Nonetheless, we will not delve into digitization challenges here, as they do not have a considerable effect on the use of integral-geometry concepts.

10.4.6 Normalization of image functionals

For presentation purposes, it is convenient to introduce the following normalized quantities:

$$\widetilde{U} = \frac{U}{L^2}, \quad \widetilde{P} = \frac{P}{LN^{1/2}}, \quad \widetilde{\chi} = \frac{\chi}{N}; \quad d = 2 \tag{10.28a}$$

and

$$\widetilde{V} = \frac{V}{L^3}, \quad \widetilde{S} = \frac{S}{L^2N^{1/3}}, \quad \widetilde{B} = \frac{2B}{LN^{2/3}}, \quad \widetilde{\chi} = \frac{\chi}{N}; \quad d = 3 \tag{10.28b}$$

where L denotes the linear size of the square (cube) and N denotes the number of germs.

10.5 Topology of periodic porous structures

In this section, we utilize MIA on more intricate surfaces. These examples aim to demonstrate the fundamental analysis approach described in Section 10.4. Specifically, we focus on calculating the Euler characteristic of a range of minimal surfaces.

A *minimal surface* in $\mathbb{R}^3$ is defined as a surface ∂A, for which the mean curvature $H = (1/R_1 + 1/R_2)/2$ is zero at each of its points, where R_1 and R_2 are the two principal radii of curvature. Hence the Gaussian curvature $G = 1/R_1R_2$ is always nonpositive. For every closed circuit on the surface, the area is a minimum. We will consider the triply periodic minimal surfaces (TPMS), minimal surfaces that are periodic in three independent directions. Structures related to triply periodic minimal surfaces may form spontaneously in materials as well as in biological systems [30]. Examples may be found in various crystal structures [31–35], lipid-containing systems [36–41], polymers [42–57], skeletal elements [58,59], and cell membranes [60]. Recently triply periodic minimal surfaces also became of interest in the analysis of the relations between the geometry and topology of the surface and surface diffusion [61–63] in *dense and porous material* systems.

A TPMS is either free of self-intersections or may intersect itself in a more or less complicated way. Each triply periodic minimal surface without self-intersections is two-sided and subdivides $\mathbb{R}^3$ into two infinite, connected but disjunct regions. These two regions, or labyrinths, are not simply connected. They interpenetrate each other in a complicated way. The two labyrinths may differ in shape, or they may be congruent, i.e., there exist symmetry operations mapping one labyrinth onto the other.

In the latter case, the surface is called a balance surface [64]. Balance surfaces divide space into two labyrinths with equal volume fractions. The symmetry of a balanced surface is described by a group-subgroup pair $\mathscr{H} / \mathscr{I}$ of space groups, which contains all isometries of $\mathbb{R}^3$ that map the surface onto itself. An isometry of $\mathscr{H}$ maps each side of the surface and each labyrinth either onto itself or onto the other side and the other labyrinth [64]. $\mathscr{I}$ contains only those isometries which map each side of the surface and each labyrinth onto itself. If the two sides of a balanced surface are "colored" so that they are symmetrically distinct, black-white space groups instead of the group-subgroup pairs with index 2 may be used to describe its symmetry [64]. In this case, the surface is called oriented. Nonbalance surfaces have $\mathscr{H} = \mathscr{I}$ and divide space into two labyrinths with unequal volume fractions.

The topology of a periodic minimal surface can be characterized by means of the Euler characteristic χ, which is related to the genus g of the surface by means of $g = 1 - \chi(\partial A)/2$ (see Section 10.3). A finite surface of genus g is the topological equivalent of a sphere with g handles.

In this context, a periodic minimal surface possesses an infinite Euler characteristic and genus. Consequently, to describe the topology of a periodic minimal surface, the genus and Euler characteristic are determined based on a unit cell. There are two standard types of unit cells: the lattice fundamental region and the crystallographic cell [65]. The lattice fundamental region is the smallest area of the surface that can replicate the entire surface through the translation of that unit cell alone. The crystallographic cell, on the other hand, is the smallest cube that generates space based on the lattice and may encompass multiple lattice fundamental regions. We will present our findings using the crystallographic cell, which will be referred to as the unit cell henceforth.

Noteworthy, the topology of a periodic minimal surface can be determined in different ways: (1) by means of the genus calculated by making use of the labyrinth graphs [66,67], (2) by means of the Euler characteristic determined with the aid of any tiling on the surface [68], (3) by means of the genus computed making use of the flat points of the surface [67], (4) by means of the simplex decomposition method [69], (5) by means of the Euler characteristic obtained with MIA.

To exemplify the advantages of MIA, we present only an analysis of the topology via the computation of the Euler characteristics and the geometry of the P (primitive), the D (double diamond), and the G (gyroid) surfaces by means of integral geometry-based MIA [29,66,71,72]. In particular, gyroidal structures in polymeric systems draw our attention as templates of *ordered periodic metallic nanofoams* [70,73]; see Figs. 10.10 and 10.11. Films of the supramolecular complex with gyroid morphology were made by dissolving the diblock copolymer of polystyrene and poly(4-vinylpyridine) (PS-b-P4VP, Polymer Source Inc.) and 3-pentadecylphenol (PDP, Aldrich 98 wt.% purity) in chloroform, followed by solvent annealing in a saturated chloroform atmosphere. For specific details, we refer to Refs. [70,73].

In addition, the same MIA analyses can be applied in characterizing the topology of the minimal balance surfaces S [64], C(P) [74], C(Y) [64], ±Y

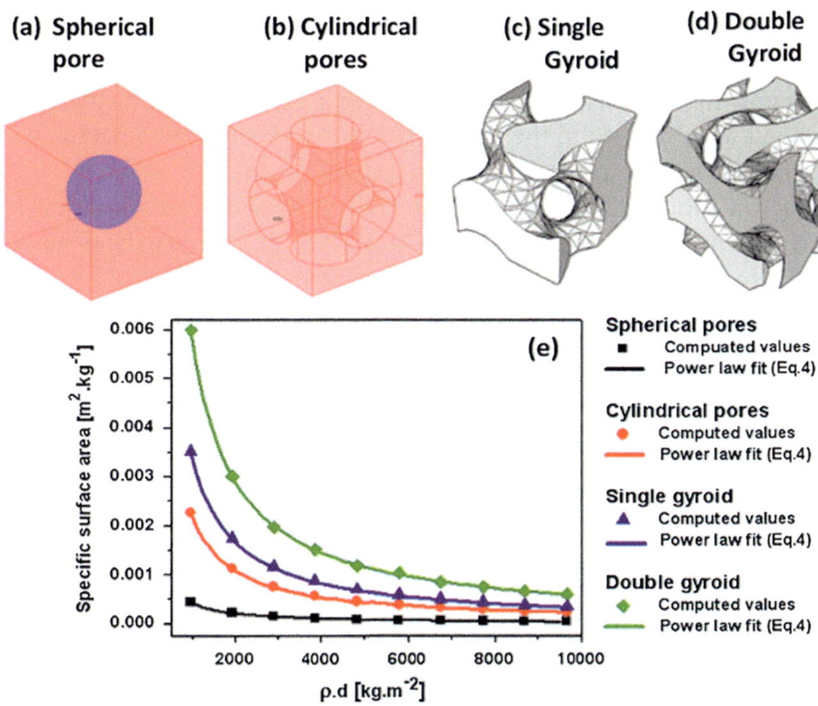

Figure 10.10 Four types of regular structures in porous metals: (a) Spherical voids, (b) cylindrical voids, (c) single gyroid, and (d) double gyroid structures. (e) For each type of structure, computed values for its specific surface area (black squares, red circles, blue triangles, and green rhombus) are in good agreement with predictions from the derived model (black, red, blue, and green lines) as reproduced with permission of Ref. [70].

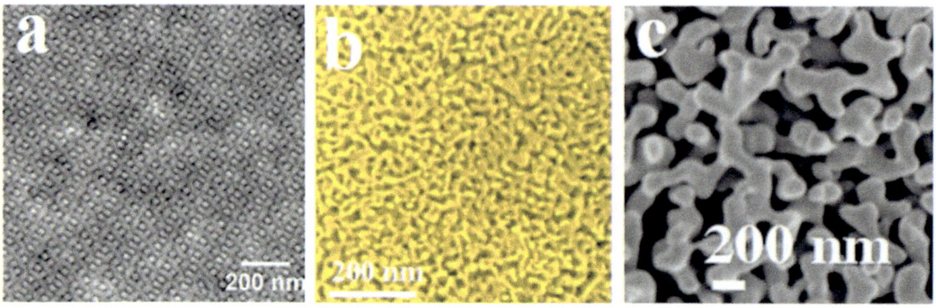

Figure 10.11 Electron micrographs showing the microstructure of synthesized nanoporous materials. (a) Transmission electron micrograph of an ordered PS-b-P4VP(PDP)x complex, a gyroid structure. This ordered morphology was made by block copolymer self-assembly. (b) Scanning electron micrograph of nanoporous gold made by the selective dissolution of silver from a gold–silver alloy. (c) Coarsened nanoporous gold ligaments after postannealing treatments. *Reproduced with permission of Ref. [70].*

[64], C(±Y) [75], and the minimal nonbalance surfaces I-WP [66] and F-RD [66].

These structures serve as a good test case for integral-geometry-based MIA because, compared to the P, D, and G surfaces, these minimal surfaces typically have a more complicated structure per unit cell.

Periodic surfaces, such as equipotential and Fermi surfaces, can be approximated by periodic *nodal surfaces* [31,34,76,77]. These can be obtained from the roots of the series:

$$\sum_{hkl}|F(hkl)|\cos(2\pi(hx+ky+lz)-\alpha_{hkl})=0, \tag{10.29}$$

where α_{hkl} and $F(hkl)$ denote a phase shift and the structure factor, respectively. Truncating the series to the leading term determines the principal nodal surfaces of a given symmetry [78].

The nodal P surface is given by

$$\cos X+\cos Y+\cos Z=0, \tag{10.30}$$

where $X=2\pi x/L_0$, $Y=2\pi y/L_0$, $Z=2\pi z/L_0$, and L_0 denotes the length of the crystallographic unit cell. The nodal primitive P surface is shown in Fig. 10.12a together with the nodal double diamond D surface in Fig. 10.12b. The latter is defined by

$$\begin{aligned}&\sin X\sin Y\sin Z+\sin X\cos Y\cos Z+\cos X\sin Y\cos Z\\&\quad+\cos X\cos Y\sin Z\\&\quad=0\end{aligned} \tag{10.31}$$

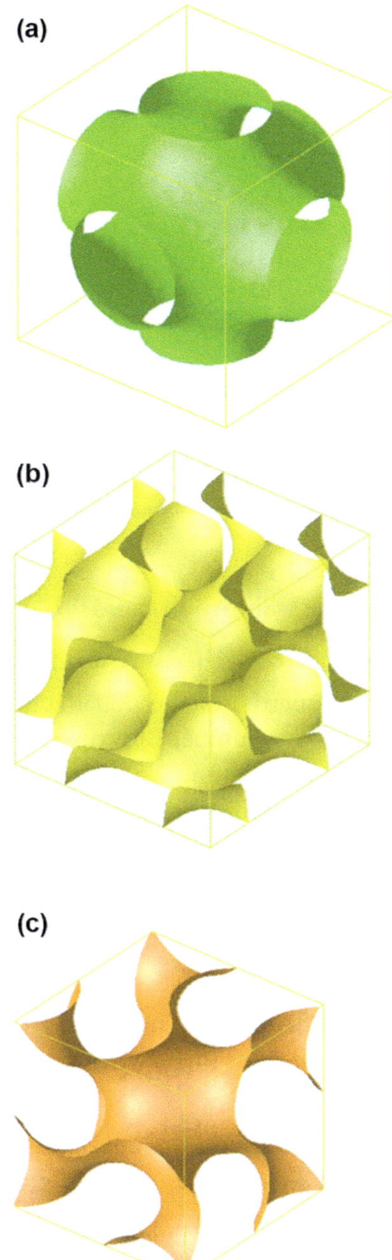

Figure 10.12 Unit cube for the nodal primitive P surface (a) and the nodal double diamond D surface (b). The surfaces are generated from Eqs. (10.30) and (10.31), respectively. (c) Unit cube for the nodal gyroid G surface. The surface is generated from Eq. (10.32).

Table 10.4 Minimal surfaces, group-subgroup pairs, and the Euler characteristic.

Minimal surface	$\mathcal{H}$	$\mathcal{J}$	$\chi^{(a)}(\partial A)$	$\chi^{(b)}(\partial A)$	$\chi^{MIA}(\partial A)$
P	$Im\bar{3}m$	$Pm\bar{3}m$	−4	−4	−4
D	$Pn\bar{3}m$	$Fd\bar{3}m$	−4	−16	−16
G	$Ia\bar{3}d$	$I4_132$	−4	−8	−8
S	$Ia\bar{3}d$	$I\bar{4}3d$	−20	−40	−40
C(P)	$Im\bar{3}m$	$Pm\bar{3}m$	−16	−16	−16
C(Y)	$I4_132$	$P4_332$	−24	−24	−24
$^{\pm}Y$	$Ia\bar{3}$	$Pa\bar{3}$	−40	—	−40
$C(^{\pm}Y)$	$Ia\bar{3}$	$Pa\bar{3}$	−24	—	−24
I-WP	$Im\bar{3}m$	$Im\bar{3}m$	−6	−12	−12
F-RD	$Fm\bar{3}m$	$Fm\bar{3}m$	−10	−40	−40

$\chi^{(a)}(\partial A)$ denotes the Euler characteristic per lattice fundamental region of the minimal surfaces obtained by means of labyrinth graphs and surface tiling [68], $\chi^{(b)}(\partial A)$ denotes the Euler characteristic per unit cell of the nodal approximations calculated by triangulating the surface with the marching cube algorithm [79], and $\chi^{MIA}(\partial A) = 2\chi^{MIA}(\partial A)$ denotes the Euler characteristic per unit cell of the nodal approximations as obtained by MIA.

The nodal gyroid surface is shown in Fig. 10.12c and is defined by

$$\sin X \cos Y + \sin Y \cos Z + \cos X \sin Z = 0. \tag{10.32}$$

It is known that the properties of the nodal approximations can differ considerably depending on the number of terms in the Fourier series and vary considerably for different structures [35,79,80]. Table 10.4 gives the results for the Euler characteristic of the nodal surfaces or one unit cell of length $L_0 = 128$ together with the numbers found in literature [68,79]. As seen from Table 10.4, the values calculated using the integral-geometry approach are in good agreement with the numbers found in literature.

In conclusion, integral-geometry-based MIA is a convenient tool for studying the topology of challenging surfaces such as some TPMS. In particular, to study the topology of the TPMS, MIA does not require the use of labyrinth graphs or surface tiling [68].

10.6 Topology of aperiodic porous structures

Metal foams have recently gained significant attention within the materials science field, although these materials have already been around for nearly 50 years [81]. Currently, metal foams with porosity levels reaching 95% are being investigated for applications that demand high specific stiffness and strength, substantial mechanical energy absorption, and heat

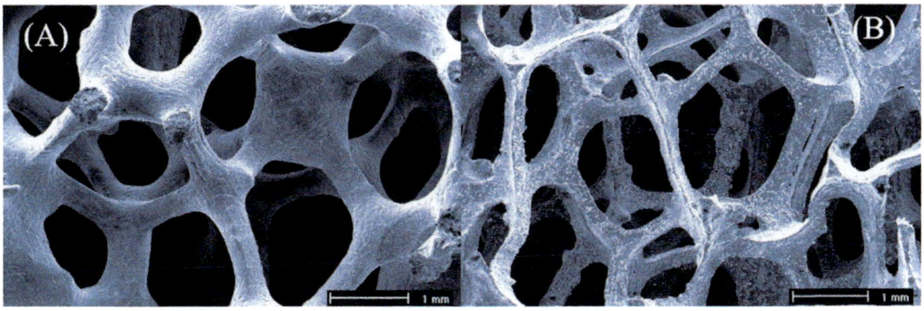

Figure 10.13 (a) SE-image of a high-density sample (10%). (b) SE image of a low-density sample (3%). Material: Duocel open-cell aluminum foam (alloy AA6101). *Reproduced with permission of Ref. [82].*

exchange functionality. Multiple studies on open-cell foams indicate that bending primarily governs both stiffness and strength, with stiffness scaling with ρ^2 and the latter with $\rho^{3/2}$ (where ρ represents the relative density). In Fig. 10.13, the bending of a strut in an open-cell foam (Duocel) is illustrated at both an early and a later stage, as observed through in-situ tensile testing.

The mechanical behavior of closed-cell foams is more complex because it involves not only the deformation of cell faces but also the edges for a thorough understanding. In metallic foams, both size effects [83] and the topology are essential factors. However, there are only a limited number of studies aimed at exploring the topological factors that influence the deformation mechanisms by examining the rigidity of frameworks made up of struts [84–86]. While commercially available metal foams typically feature random microstructures, it is also possible to create microstructures with periodic architectures that possess superior properties compared to their stochastic counterparts [87].

For materials scientists, it is of interest to investigate the topology of the metal porous structures without destroying the sample. Using X-ray tomographic (CT) scans makes it possible to visualize the interior of a porous system in a nondestructive manner and to produce a 3D image of the sample.

Information regarding the morphological characteristics can be derived from calculating the Minkowski functionals for these 3D CT scans. In this section, we begin by detailing how we compute the 3D Minkowski functionals from the CT scans. We then utilize this method on two different types of aluminum foam samples. We show that applying morphological image analysis techniques to examine the 3D CT scans provides insights into the morphological features and may assist in classifying the metallic structures.

10.6.1 Computation of 3D Minkowski functionals

High-resolution CT scanners can generate large datasets of 2D slices, usually around 1 MB each. These 2D slices are utilized to create a 3D image. Our 3D images consist of cubes extracted from the comprehensive CT image. Prior to calculating the Minkowski functionals, we apply a threshold to the 3D image. The application designed to compute Minkowski functionals for the 3D images keeps four slices in memory at a time. Essentially, we employ the same approach as outlined in Section 10.3. Before incorporating a pixel, we verify whether the pixel to be added has active nearest or next-nearest neighbors and tally the number of faces, edges, and vertices accordingly. At the beginning of the computation, the first three slices are entirely empty (all pixels are white), while the fourth slice holds the pixels from the initial slice of the CT scan. We remove the pixels from the fourth slice (one by one, from left to right and from front to back) and add them to the second slice after checking its nearest neighbors and/or next nearest neighbors in all directions (the bottom direction is the first slice and the top direction is the third slice) and counting the number of faces, edges, and vertices. We calculate the changes to V, S, B, and χ and proceed. If the fourth slice is emptied, we make a copy of the second slice in the first slice and load the second CT slice in the fourth slice. We now proceed in the same way as for the first step. The number of arithmetic operations required to calculate the Minkowski functionals scales linearly with the number of active pixels in the image. The memory needed equals four times the memory required to store one CT slice. The computational and memory demands can be slightly reduced. A computer code was written for more general applications than the one we need here. The program can handle the situation in which we remove one pixel from an image of which we already calculated the Minkowski functionals and compute the change in the Minkowski functionals. This requires the check of all nearest and next-nearest neighbors (26 in total). In the CT case, we build up the image by adding pixel by pixel. Then only half of the neighbors need to be checked. Moreover, only storage for three slices is required.

To calculate the Minkowski functionals in a more efficient manner, the following procedure can also be employed. When the computation begins, the initial two slices are entirely empty (all pixels are white), while the third slice contains the pixels from the first CT scan slice. We sequentially remove pixels from the third slice (one at a time, moving from left to right and from front to back) and transfer them to the second slice, checking their nearest and/or next nearest neighbors to the left, front, and bottom (the first slice) while counting the number of faces, edges, and vertices. We calculate the changes to V, S, B, and χ and proceed. If the third slice is emptied, we make a copy of the second slice in the first slice and load the second CT slice in the third slice. We now proceed in the same way as for the first step and continue until all CT slices are processed.

10.6.2 Examples

An illustration of a 2D CT scan of closed-cell aluminum foam (left) and open-cell aluminum foam (right) is presented in Fig. 10.14. We utilized hundreds of these 2D images to reconstruct a 3D representation of both closed-cell and open-cell aluminum foam. In Fig. 10.15, 3D visualizations of the two distinct aluminum foam samples are shown. The image on the left depicts a closed-cell foam, whereas the one on the right exhibits a foam with a more porous structure. To analyze the Minkowski functionals for these aluminum foams, we initially applied a threshold to the cubic image. For the images in question, only very small thresholds are significant. Using medium to large thresholds ($q > 50$) results in the loss of small sections and entirely disconnected structures within the image. At these higher thresholds, the resulting image fails to accurately depict the genuine foam structure.

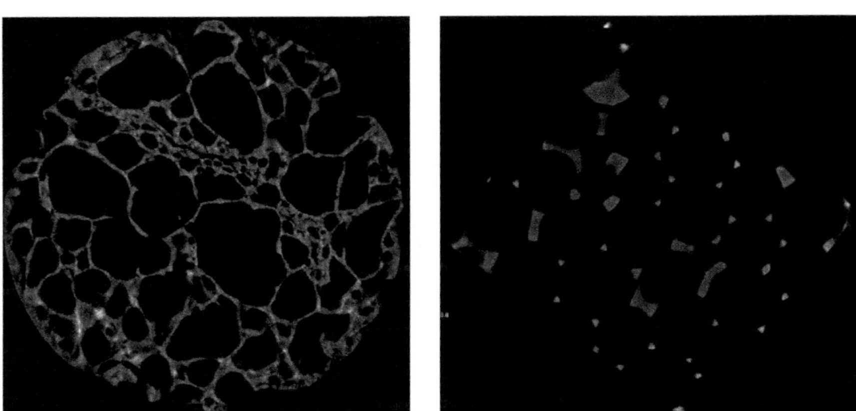

Figure 10.14 2D computer X-ray tomography image of a closed-cell (left) and open-cell (right) aluminum porous structure.

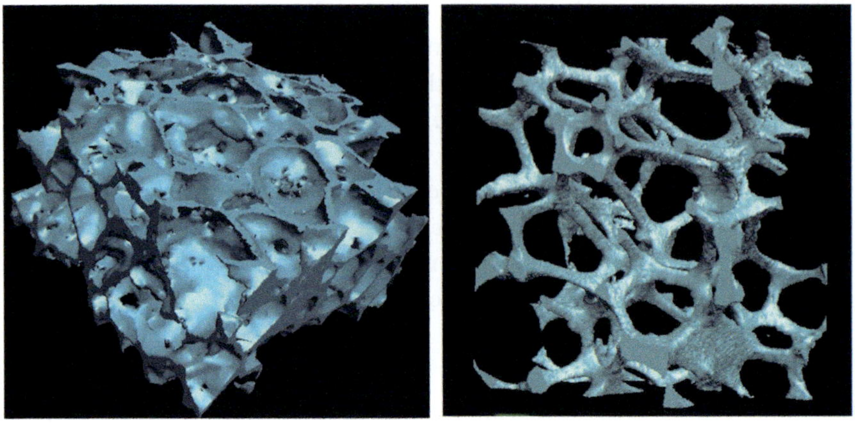

Figure 10.15 3D computer X-ray tomography images of two aluminum foams of different types. Left: Closed-cell aluminum foam; right: Open-cell aluminum foam.

In Figs. 10.16 and 10.17, we depict the Minkowski functionals as a function of the threshold q for the closed-cell and open-cell structures, respectively. For the analysis of the foam structures, we only consider $0 < q < 50$ because for larger q, the image is almost completely empty.

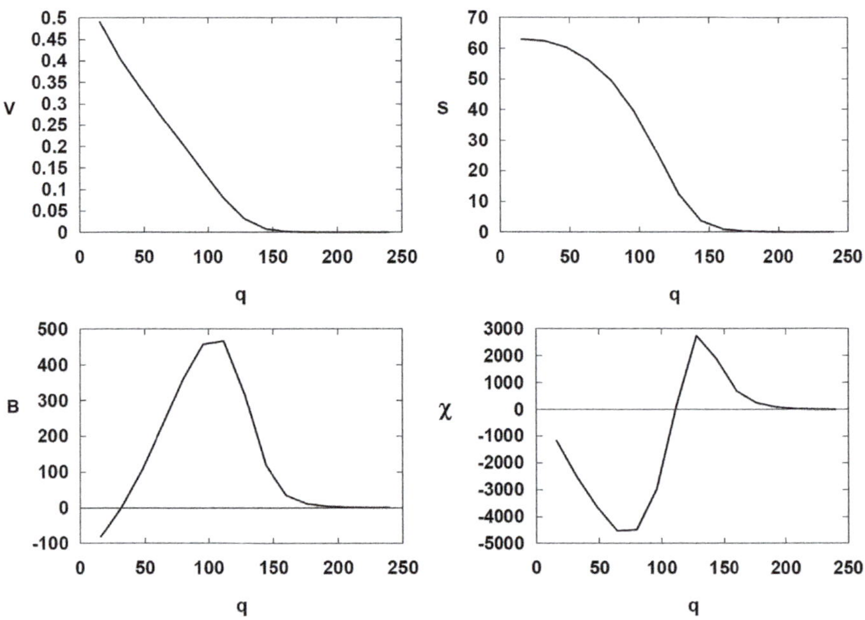

Figure 10.16 Minkowski functionals as a function of the threshold q obtained from a 3D X-ray computer tomography image of a dense aluminum foam. *With permission of Ref. [4].*

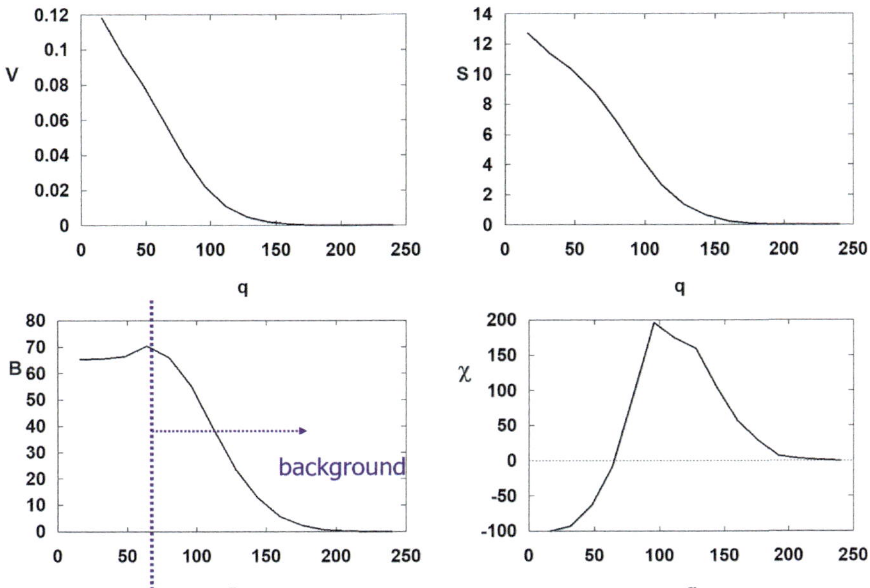

Figure 10.17 Minkowski functionals as a function of the threshold q obtained from a 3D X-ray computer tomography image of an open-cell aluminum foam. *With permission of Ref. [4].*

The volume covered by the closed-cell foam is significantly larger (four times) compared to the foam with a more open configuration. The surface areas of the two types of foam exhibit a similar characteristic. This indicates that the aluminum foam with an open structure contains large voids (cavities and tunnels) filled with air, while the closed-cell aluminum foam comprises many small voids and tunnels that are also filled with air. This observation can be further illustrated by the structural images in Fig. 10.15. In the case of the open-cell aluminum foam, the average breadth remains positive and fairly consistent for $0 < q < 50$. In the case of the closed-cell foam, the mean breadth changes from negative to positive values for $0 < q < 50$. This indicates that in the closed-cell foam, more tunnels are present. For both foams, the Euler characteristic is negative, which means that on average, the surface is hyperbolic and contains a lot of tunnels. For the foam with the closed-cell structure, the Euler characteristic is much more negative than the Euler characteristic for the foam with the more open structure. The more negative χ becomes the more tunnels in the structure and the more complex the structure is [88].

The analysis of the 3D Minkowski functionals clearly indicates that the two aluminum foams have a different internal structure. A more detailed morphological characterization of the foams can be made by investigating local instead of averaged Minkowski functionals. Additional information can also be obtained from the computation of the 2D Minkowski functionals of various slices.

It is noteworthy that when comparing closed-cell and open-cell porous materials, quantitative data are obtained through MIA analyses. For instance, besides the Euler characteristics $\chi(K)$ also the surface area *S(K)* and *B(K)*, the mean breadth appears in the formulation for convex sets, e.g.,

$$V(K_r) \propto S(K)r + 2\pi B(K)r^2 + \frac{4}{3}\pi\chi(K)r^3. \tag{10.33}$$

As a consequence, the *S(K),* being proportional to the number of faces, should be significantly higher in closed-cell compared to open-cell porous materials, as is indeed confirmed in Fig. 10.16 (closed cell) and Fig. 10.17 (open-cell), i.e., 60 and 10, respectively. Consistently, the mean breadth *B(K)* in closed-cell systems should be much smaller, since it is based on calculating, see also Eq. (10.25):

$$2B(K) = n_{\text{edges}} - 2n_{\text{faces}} + 3n_{\text{cubes}}, \tag{10.34}$$

again, in accordance with Figs. 10.16 and 10.17, −20 (closed-cell) and +70 (open-cell), respectively. Also, $\chi(K)$ of closed-cell systems should be much smaller than open-cell systems due to

$$\chi(K) = \sum_{\nu=0}^{d} (-1)^{\nu} n_{\nu} = n_{vertices} - n_{edges} + n_{faces} - n_{cubes} \tag{10.35}$$

and it is pointing to cavities connected by larger torus-like holes, with both *B(K)* and $\chi(K)$ strongly negative in comparison with open-cell structures (see Figs. 10.16 and 10.17).

It is clear that for effectively utilizing porous systems in mechanical engineering, understanding their structure under mechanical stress is crucial. Cracks can propagate through void coalescence, making it essential to have insights into the failure mechanisms within the cell walls. Cell wall necking has been identified as the primary failure mode, and various studies across a diverse range of both open and closed-cell foams, produced using different manufacturing techniques, have demonstrated that microstructure significantly impacts failure. However, conducting a comprehensive quantitative analysis of the morphology, beyond merely qualitative assessments

Figure 10.18 Left: CT scan; Right: 3D X-ray computer tomography reconstructed image of deformed closed-cell aluminum porous system.

through electron microscopy, is more challenging. Specifically, the morphological image analysis method presented here can accomplish this objective, providing quantitative data to elucidate the relationships between structure and properties.

To illustrate this point in further detail. Fig. 10.18 shows the CT scan and reconstructed 3D image of the same closed-cell aluminum system, but now plastically deformed under compression. Clearly it is very hard to get any data out of the 3D reconstructed image: quantitatively yes, that is to say only to a certain extent, but definitely not in a quantitative manner with 'physical numbers. MIA provides an escape route through these data, e.g., of Euler characteristics $\chi(K)$ and mean breadth *B(K)*, as displayed in Figs. 10.19 and 10.20, respectively. Clearly, deformation leads to a strongly negative $\chi(K)$ as well as negative *B(K)*, pointing to increasing small cavities connected by torus like holes (see also Fig. 10.16, nondeformed Al closed-cell structures).

In this chapter, we have demonstrated that the morphological image analysis technique is useful in giving information on the 2D/3D morphological properties of metal porous structures. Although the amount of data obtained by CT scans is huge, the memory and computation time requirements for the calculation of the Minkowski functionals are rather low. In order to study the morphological properties of various metal porous systems in more detail, more samples can be analyzed and easily compared with each other.

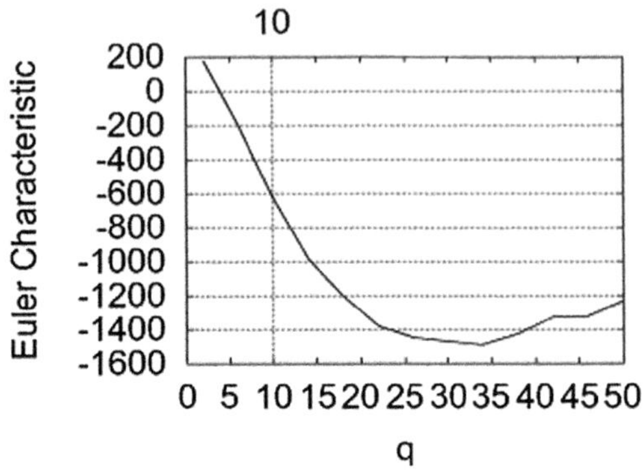

Figure 10.19 Minkowski functional Euler number χ as a function of the threshold q obtained from a 3D X-ray computer tomography image of a deformed closed-cell aluminum porous system.

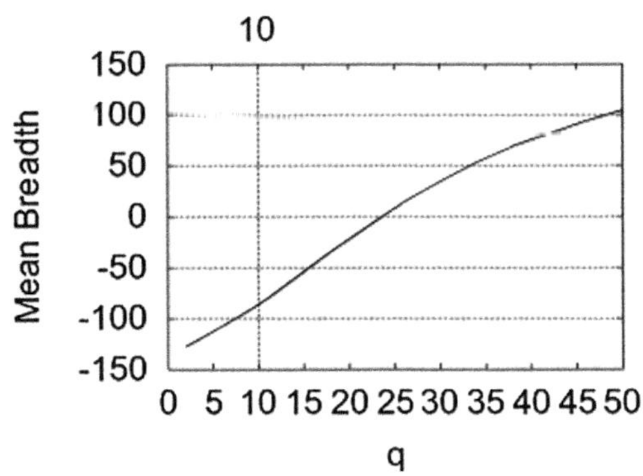

Figure 10.20 Minkowski functional mean breadth (B) as a function of the threshold q obtained from a 3D X-ray computer tomography image of a deformed closed-cell aluminum porous system.

In principle, the analysis method applies both to X-ray scanning and also to scanning electron microscopy. Of course, the spatial resolutions differ a lot, going down from say 10 μm length scale in X-ray CT (80 kV, 100 μA , spot size 10 μm) to say 1 nm in scanning electron microscopy (SEM).

In X-ray computer tomography, a polychromatic X-ray source illuminates the object and a 2D detector collects magnified projection images. Based on angular views acquired while the object rotates, a stack of virtual cross-section slices is computed.

3D tomography is more straightforward with X-ray CT, but we have also established a calibration-free approach for quantitatively reconstructing surface topography in SEM that we have named TopoSEM. In this method, a series of images captured at various tilt angles are analyzed using digital image correlation (DIC). The method strength lies in the fact that detailed knowledge about the rotation characteristics (both vector and magnitude) is unnecessary. Consequently, a significant benefit is that complicated calibrations of the measuring instruments are eliminated, and the approach discussed in Refs. [89–91]. TopoSEM provides the topographical information hidden in the contrast of the SEM pictures, avoiding the necessity of calibrations and other complex geometrical considerations.

10.7 Conclusions

To characterize the morphological aspects of a porous materials system, one must consider both topological and geometrical descriptors to describe the pattern connectivity, content, and shape. The approach discussed here, known as morphological image analysis (MIA), relies on the principles of integral and differential geometry to reveal the geometrical and topological structures of point sets in Euclidean space. Integral geometry provides these descriptors through the Minkowski functionals. To facilitate understanding of the process, the term image functional refers to a function that assigns a numerical value to a (black-and-white) image, indicating the measurement of a specific property of that image.

Most image functionals possess the property of additivity, meaning all features present in the image are accounted for only once. Any morphological measurement can be expressed as a linear combination of the $d+1$ Minkowski functionals in d dimensions (which are additive, motion invariant, and continuous); this is applicable to any geometrical object. MIA, based on integral geometry, employs a collection of additive image functionals that assess the shape and connectivity of various objects.

In practice, the method of image analysis presented is based on Minkowski functionals in *pixel* representations, realizing that each pixel in a 2D(3D) image is a convex set. The calculation of the Minkowski functionals amounts to counting vertices and edges in a pixel representation of CT images of metal porous systems, e.g., following the recipe of making a scan of the image pixel-by-pixel, compute per pixel the change in $2D$ of $\Delta A, \Delta\ L$ and $\Delta\chi$ by calculating the changes in simplexes $\Delta n_0, \Delta n_1, \Delta n_2$ where $n_0 =$ # vertices, $n_1 =$ # edges, $n_2 =$ # squares. Add these Δ changes to the total of A, L and χ. An extremely useful quantity to describe the porous system is the Euler characteristic χ , which is motion invariant and identical to the one defined in algebraic topology, where $\chi(K)$ is directly related to the number of cavities and torus-like holes (see Eq. 10.35).

The appeal of MIA lies in its straightforward implementation, supported by a robust mathematical foundation. One notable characteristic of MIA is the significant disparity between the intricate mathematics involved and the simplicity of its implementation and use. MIA does not necessitate a regular surface, nor does it require the introduction of labyrinth graphs or surface tiling for derivative calculations. Instead, MIA can be applied directly to the digitized representation of patterns, can be executed with just a few lines of code, is computationally efficient, and is practical to use. As a result, it serves as an attractive toolset for analyzing geometrical objects and patterns in porous systems.

Appendix 10.A: Noise and artifacts

Electron microscopy, both in scanning and transmission mode, is used in analyses of geometric and topological characteristics of porous materials. The signals produced by an SEM include secondary and back-scattered electrons (SE&BSE), characteristic X-rays, cathodoluminescence (light), specimen current, and transmitted electrons. The SE&BSE detector is the most common one in SEM.

As far as structural analyses are concerned, transmission electron backscatter diffraction (coined t-EBSD) turns out to be a rather appealing method in this field for analyzing microstructures, which relies on capturing Kikuchi diffraction patterns produced by electrons that have passed through an SEM. This advancement, which offers enhanced resolution compared to traditional electron backscatter diffraction (EBSD) [92], was employed in our earlier work to study the creation of nanoporous gold through the free dealloying of these foils [93,94]. Notably, it was observed that the original <001> and <111> fiber textures in the $Au_{54}Ag_{46}$ master alloy were strengthened following the etching process, while the residual strain, assessed by the average misorientation within grains, showed a decline. In fact, the development of low-angle grain and twin boundaries at the expense of high-angle grain boundaries suggested that grain boundaries function as nucleation sites for the etching process, producing highly porous materials on the submicrometer and nanometer scale [94–96].

So far, as regards the analysis of topology using MIA, we have not discussed specifically the effects of noise in SEM characterization and have assumed that the digitized images are entirely free from noise and other artifacts that could affect the outcomes of geometric and topological analyses. In the context of porous media, various issues may arise due to the charging of the electron beam when utilizing electron microscopy and generally because of mechanical or electrical noise.

This can result in blurry edges and often causes the edges to exhibit different brightness than the centers, complicating the process of applying a threshold to the image. Problems frequently arise when analyzing magnetic materials or due to the influence of stray magnetic fields. The thermal noise from the SEM beam scanning system is recognized as the primary cause of the "artificial" displacements seen between successive SEM scans conducted under identical conditions, as analyzed by digital speckle correlation [97]. In fact, the accuracy at both pixel and subpixel levels typically declines with higher SEM magnification, resulting in a limitation on the potential for downscaling. It has been shown that SEM images captured at a magnification of 40k with a pixel size of 2.8 nm exhibit a standard deviation of approximately 0.1 pixel, leading to an accuracy of around 0.3 nm for each measured point [98]. Generally, various sources of drift are reported, including drift in the electron beam positioning during scanning, sample stage drift, errors in repositioning after stage tilting for ion milling, and surface charging of the object.

For optimum scanning, different parameters can be used depending on the specimen character or the quality of the desired image. Some of these parameters are depicted in Fig. A.10.1, which shows the schematic of image formation in an SEM. Each pixel in the SEM image corresponds to each

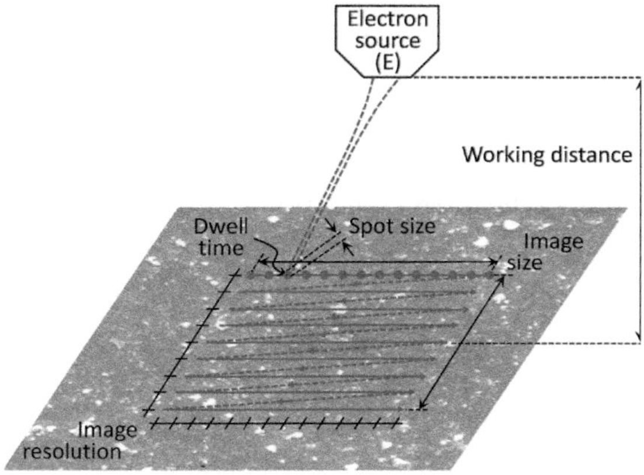

Figure A.10.1 Schematic of the formation of an image in a scanning electron microscope. *Reproduced with permission of Ref. [99].*

position of the electron beam on the sample. The distances between the spots in horizontal and vertical directions are the same, as well as the size of the pixels in both of these directions. The ratio between the image pixel size and the distance between two spots on the sample defines the magnification of the image.

Currently, the field of view is preferred to provide information about the magnification. Nevertheless, there are other options that influence the final acquisition of an image. For instance, an additional delay time could be required before starting the line scan in order to synchronize the scanning process with the frequency of the power line. This feature is referred to as synchronization, and depending on the microscope and the dwell time (line synchronization), it can be toggled on or off. When an image is captured after a single scan, it consists of only one frame. The latest SEM models provide the capability to gather images by increasing the number of frames (number of scans), whereby the final image is created by integrating signal intensities for each pixel from the individual frames. This feature is known as integration. Furthermore, advanced SEMs can also perform synchronization between frames.

Various research studies in the literature indicate that the stability of imaging is influenced by the specific SEM systems in use, across a broad range of magnifications from 200 to 10k. These largely qualitative investigations reveal that each of these microscopes experiences an image shift at random locations during SEM imaging, suggesting that image inaccuracies are an inherent characteristic of this process. In Ref. [99], we introduced a more quantitative statistical approach to examine the instabilities that may arise during imaging in SEM. This technique is based on the correlation of digital images and can be used at different length scales.

To address the issues caused by noise from a mathematical perspective, the specialized method of MIP & MIA morphological image processing as demonstrated in Section 10.4 was designed to evaluate the distances between the centers of ligaments and pores. In this context, two pairs of morphological operators are crucial. These operators are referred to as dilating, eroding, opening, and closing. It is evident that the Minkowski addition (refer to Section 10.3.2) increases the size of the resultant set, while Minkowski subtraction reduces it. These properties are used to define two new operators: dilation and erosion.

First, *Dilation* is defined as:

$$\boldsymbol{D}(\boldsymbol{A},\boldsymbol{B}) = \boldsymbol{A} \oplus \boldsymbol{B} \tag{A.10.1}$$

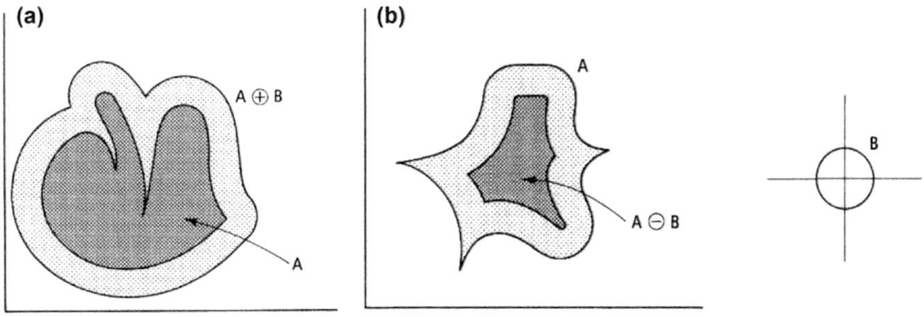

Figure A.10.2 A dilation *(a)* and an erosion *(b)* of object *A* by structuring element *B*. *Reproduced with permission of Ref. [100].*

This looks identical to the Minkowski addition, and actually, it almost is (see Section 10.3.2.). There is a subtle difference, though; the term *Dilation* is used when one of the sets (A) is much larger than the other (B). The smaller element is then called the *structuring element.*

Second, *Erosion*, on the other hand, makes object A smaller by using a structuring element B. The definition of erosion is:

$$\boldsymbol{E}(\boldsymbol{A},\boldsymbol{B}) = \boldsymbol{A}(-)(-\boldsymbol{B}). \tag{A.10.2}$$

In Fig. A.10.2a, Minkowski addition *(a)* and a Minkowski subtraction *(b)* of object A by structuring element B are shown. Dilation "rolls" the structuring element around the border of the various objects, creating new objects that contain the border plus the structuring element. Eroding creates new objects for which the structuring element is fully contained in the original object. Thus, dilation tends to expand the subject with the structuring element, while erosion tends to shrink the subject.

Since dilation and erosion have been properly defined, we can now continue to define more complex operators. The most important and commonly used are *Opening* and *Closing*.

For an image A and structuring element B *Opening* is defined as

$$\boldsymbol{O}(\boldsymbol{A},\boldsymbol{B}) = \boldsymbol{D}(\boldsymbol{E}(\boldsymbol{A},\boldsymbol{B}),\boldsymbol{B}) \tag{A.10.3}$$

So, it is simply an erosion followed by a dilation. In Fig. A.10.3 is shown for a noisy image A that first is eroded by structuring element B. The result is then dilated to get the outcome of the open operator. From Fig. A.10.3 it can be seen that $O(A, B)$ as defined by Eq. (A.10.3) is a sub-image of A. The noise around the main structure has been effectively removed.

Closing is defined as a dilation followed by an erosion, i.e.,

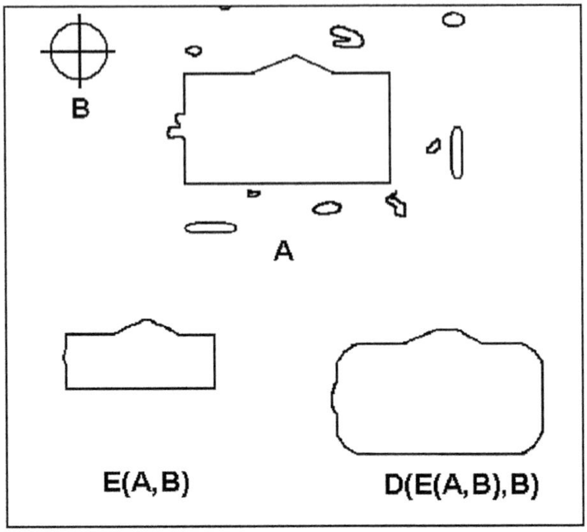

Figure A.10.3 Opening of image *A* by structuring element *B* shown as an erosion and dilation.

$$C(A,B) = E(D(A,B),B) \quad \text{(A.10.4)}$$

Intuitively opening an image is obtained by "rolling" the structuring element about the inside of the image while, closing is obtained by rolling the structuring element around the outside of the image. This is shown in Fig. A.10.4 in the case of Closing where a rectangular object A, with some scattering around an inside, first is dilated and subsequently is eroded with structuring element B. It is concluded that *Opening* and *Closing* an image with an appropriate structuring element can free an image from noise. The size of the structuring elements, defined in concurrence with the magnification and resolution of the SEM pictures, turned out to be crucial to the successful processing of the images.

Appendix 10.B: Algorithm

We present a method for assessing the changes in the count of open bodies in each category when a black pixel is either introduced or taken away from a selected pattern. Employing this method enables us to effectively calculate the Minkowski functionals for a given pattern, which can be done by incrementally adding black pixels to a completely white image.

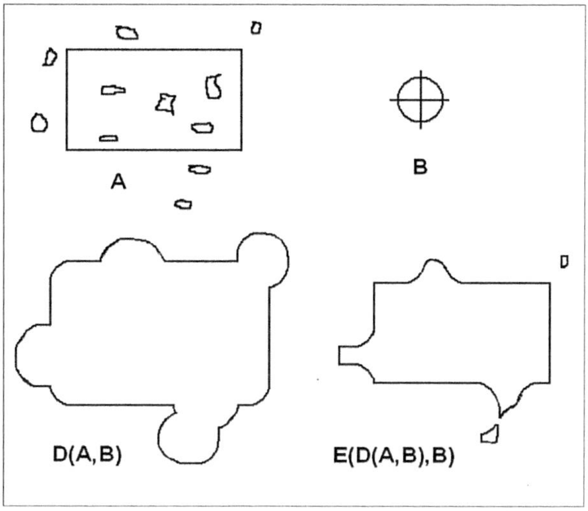

Figure A.10.4 Subsequent dilation and erosion of object *A* by structuring element *B* is closing an image.

In 2D, the number $n_2(\mathscr{P})$ of open squares building up the objects in the $L_x \times L_y$ image $\mathscr{P}(\mathbf{x}, q) = (i, j, q)$; $(i = 1, \ldots, L_x, j = 1, \ldots, L_y)$ increases (decreases) with one if one adds (removes) one black pixel at the position $\mathbf{x} = (i, j)$ to (from) the image. Therefore, if we add a black pixel,

$$\Delta n_2(\mathscr{P}) = 1 \tag{B.10.1}$$

where we introduce the symbol Δ to indicate that we compute the difference. Similarly, the change in the number of open edges, $\Delta n_1(\mathscr{P})$ is given by

$$\Delta n_1(\mathscr{P}) = \sum_{\alpha=\pm 1} [\mathscr{Q}(i+\alpha, j, q) + \mathscr{Q}(i, j+\alpha, q)], \tag{B.10.2}$$

where $\mathscr{Q}(\mathbf{x}, q) = 1 - \mathscr{P}(\mathbf{x}, q)$.

For the change in the number of vertices, $n_0(\mathscr{P})$, we find

$$\Delta n_0(\mathscr{P}) = \sum_{\alpha,\beta=\pm 1} [\mathscr{Q}(i+\alpha, j, q)\mathscr{Q}(i+\alpha, j+\beta, q)\mathscr{Q}(i, j+\beta, q)] \tag{B.10.3}$$

In 3D, the number $n_3(\mathscr{P})$ of open cubes building up the objects in the $L_x \times L_y \times L_z$ image $\mathscr{P}(\boldsymbol{x}, q) = \mathscr{P}(i, j, k, q);\ i = 1, \ldots, L_x, j = 1, \ldots, L_y, k = 1, \ldots, L_z)$ increases (decreases) with one if one adds (removes) one black pixel to (from) the image at the position $\boldsymbol{x} = (i, j, k)$,

i.e., $\Delta n_3(\mathscr{P}) = 1$. The change in $\Delta n_2(\mathscr{P})$ the number of open faces, may be computed from

$$\Delta n_2(\mathscr{P}) = \sum_{\alpha=\pm 1} [\mathscr{Q}(i+\alpha, j, k, q) + \mathscr{Q}(i, j+\alpha, k, q)\mathscr{Q}(i, j, k+\alpha, q)] \tag{B.10.4}$$

The change in $\Delta n_1(\mathscr{P})$ the number of open edges, reads

$$\begin{aligned}\Delta n_1(\mathscr{P}) = \sum_{\alpha,\beta=\pm 1} & [\,\mathscr{Q}(i+\alpha, j, k, q)\mathscr{Q}(i+\alpha, j+\beta, k, q)\mathscr{Q}(i, j+\beta, k, q)] \\ & + [\,\mathscr{Q}(i, j+\alpha, k, q)\mathscr{Q}(i, j+\alpha, k+\beta, q)\mathscr{Q}(i, j, k+\beta, q)] \\ & + [\,\mathscr{Q}(i+\alpha, j, k, q)\mathscr{Q}(i+\alpha, j, k+\beta, q)\mathscr{Q}(i, j, k+\beta, q)]\end{aligned} \tag{B.10.5}$$

For the change in the number of vertices, we find

$$\begin{aligned}\Delta n_0(\mathscr{P}) = \sum_{\alpha,\beta,\gamma=\pm 1} & \mathscr{Q}(i+\alpha, j, k, q)\mathscr{Q}(i+\alpha, j+\beta, k, q)\mathscr{Q}(i, j+\beta, k, q) \\ & \times \mathscr{Q}(i+\alpha, j, k+\gamma, q)\mathscr{Q}(i+\alpha, j+\beta, k+\gamma, q) \times \mathscr{Q}(i, j \\ & + \beta, k+\gamma, q)\mathscr{Q}(i, j, k+\gamma, q)\end{aligned} \tag{B.10.6}$$

Appendix 10.C: Programming example (Fortran 90)

In this section a computer program is displayed to compute 3D Minkowski functionals in three dimensions, volume, surface, mean curvature, and Euler. The program was first designed, developed, and published in Ref. [29] for the 2D case (area, perimeter, and Euler). In particular, for the 2D version, reference is made to Ref. [29] (Fig. C.10.1).

```
!
! Minkowski functionals 3D computes the Minkowski functionals
! (volume,surface,integral mean curvature,euler) for a 3D image,
! represented by the 1D array LATTICE(.). A pixel at (jx,jy,jz) is
! black if LATTICE(jx+Lx*(jy+Ly*jz))=1, otherwise LATTICE(jx+Lx*(jy+Ly*jz))=0.
! Here 0 < jx < Lx, 0 < jy < Ly, and 0 < jz < Lz. The array TMP(.)
! is used as work space.
! Putting FREE_BOUNDARIES = 0 returns Minkowski functionals for
! periodic boundary conditions, other values return Minkowski functionals
! for the image on in a infinitely large empty background.

      subroutine minkowski functionals(Lx,Ly,Lz,free boundaries,tmp, &
                                   lattice,volume,surface,curvature,euler)
      implicit integer (a-z)
      integer lattice(0:*),tmp(0:*)

      vol=0
      sur=0
      cur=0
      eul=0

      if(free boundaries.eq.0) then ! periodic boundary conditions
       tmp(0:Lx*Ly*Lz-1)=0
      else
       tmp(0:(Lx+2)*(Ly+2)*(Lz+2)-1)=0
      endif

      do jz=0,Lz-1
      do jy=0,Ly-1
      do jx=0,Lx-1
       i=jx+Lx*(jy+Ly*jz)
        if( lattice(i) > 0 ) then ! black pixel
          if(free boundaries.eq.0) then ! periodic boundary conditions
           call minko_3D periodic(Lx,Ly,Lz,jx,jy,jz,tmp,v,s,c,e)
           tmp(i)=1 ! can only be 0 or 1 in minko 3D periodic
          else
           call minko_3D free(Lx+2,Ly+2,Lz+2,jx+1,jy+1,jz+1,tmp,v,s,c,e)
           tmp(jx+1+(Lx+2)*(jy+1+(Ly+2)*(jz+1)))=1
          endif
         vol=vol+v
```

Figure C.10.1 Computer code (Fortran 90) for the computation of Minkowski functionals of a 3D image.

```
jyi=jy+i0
if(jyi.lt.0) then
jyi=Ly+jyi
else if(jyi.ge.Ly) then
jyi=jyi-Ly
endif
jzi=jz+i0
if(jzi.lt.0) then
jzi=Lz+jzi
else if(jzi.ge.Lz) then
jzi=jzi-Lz
endif
kc1=1-lattice(jxi+Lx*(jy+Ly*jz))
kc2=1-lattice(jx+Lx*(jyi+Ly*jz))
kc3=1-lattice(jx+Lx*(jy+Ly*jzi))
nfaces=nfaces+kc1+kc2+kc3
do j0=-1,1,2
jyj=jy+j0
if(jyj.lt.0) then
jyj=Ly+jyj
else if(jyj.ge.Ly) then
jyj=jyj-Ly
endif
jzj=jz+j0
if(jzj.lt.0) then
jzj=Lz+jzj
else if(jzj.ge.Lz) then
jzj=jzj-Lz
endif
k4=Lx*(jyj+Ly*jz)
k7=Lx*(jy+Ly*jzj)
kc7=1-lattice(jx+k7)
kc1kc4kc5=kc1*(1-lattice(jxi+k4))*(1-lattice(jx+k4))
nedges=nedges+kc1kc4kc5+kc2*(1-lattice(jx+Lx*(jyi+Ly*jzj)))*kc7 &
           +kc1*(1-lattice(jxi+k7))*kc7
if(kc1kc4kc5.ne.0) then
do k0=-1,1,2
jzk=jz+k0
if(jzk.lt.0) then
jzk=Lz+jzk
else if(jzk.ge.Lz) then
jzk=jzk-Lz
endif
```

Figure C.10.1 (*continued*).

```
    sur=sur+s
    cur=cur+c
    eul=eul+e
  endif

enddo
enddo
enddo

volume=vol
surface=sur
curvature=cur
euler=eul

end

subroutine minko_3D_periodic(Lx,Ly,Lz,jx,jy,jz,lattice, &
                              volume,surface,curv,euler3D)
implicit integer (a-z)
integer lattice(0:LX*LY*Lz-1)
parameter(              &
  volume_body=1    , &      !(a*a*a, where a is lattice displacement)
  surface_body=-6  , &      !(-6*a*a, open body)
  surface_face=2   , &      !(2*a*a, open face)
  curv_body=3      , &      !(3*a, open body)
  curv_face=-2     , &      !(-2*a, open face)
  curv_edge=1      , &      !(a, open line)
  euler3D_body=-1  , &      !(open body)
  euler3D_face=1   , &      !(open face)
  euler3D_edge=-1  , &      !(open line)
  euler3D_vertex=1)          !(vertices)

nfaces=0
nedges=0
nvert=0

do i0=-1,1,2
jxi=jx+i0
if(jxi.lt.0) then
jxi=Lx+jxi
else if(jxi.ge.Lx) then
jxi=jxi-Lx
endif
```

Figure C.10.1 (*continued*).

```
k9=Lx*(jy+Ly*jzk)
k10=Lx*(jyj+Ly*jzk)
nvert=nvert+(1-lattice(jxi+k9))*(1-lattice(jxi+k10)) &
            *(1-lattice(jx+k10))*(1-lattice(jx+k9))
enddo ! k0
endif ! kc1kc4kc5
enddo ! j0
enddo ! i0

volume=volume_body
surface=surface_body+surface_face*nfaces
curv=curv_body+curv_face*nfaces+curv_edge*nedges
euler3D=euler3D_body+euler3D_face*nfaces &
        +euler3D_edge*nedges+euler3D_vertex*nvert

return
end

subroutine minko_3D_free(Lx,Ly,Lz,jx,jy,jz,lattice, &
                           volume,surface,curvature,euler3D)

implicit integer (a-z)
integer lattice(0:LX*LY*Lz-1)
parameter(             &
  volume_body=1    , &      !(a*a*a, where a is lattice displacement)
  surface_body=-6  , &      !(-6*a*a, open body)
  surface_face=2   , &      !(2*a*a, open face)
  curv_body=3      , &      !(3*a, open body)
  curv_face=-2     , &      !(-2*a, open face)
  curv_edge=1      , &      !(a, open line)
  euler3D_body=-1  , &      !(open body)
  euler3D_face=1   , &      !(open face)
  euler3D_edge=-1  , &      !(open line)
  euler3D_vertex=1)          !(vertices)

nfaces=0
nedges=0
nvert=0

do i0=-1,1,2
jxi=jx+i0
jyi=jy+i0
jzi=jz+i0
```

Figure C.10.1 (***continued***).

```
kc1=1-lattice(jxi+Lx*(jy+Ly*jz))
kc2=1-lattice(jx+Lx*(jyi+Ly*jz))
kc3=1-lattice(jx+Lx*(jy+Ly*jzi))
nfaces=nfaces+kc1+kc2+kc3
do j0=-1,1,2
jyj=jy+j0
jzj=jz+j0
k4=Lx*(jyj+Ly*jz)
k7=Lx*(jy+Ly*jzj)
kc7=1-lattice(jx+k7)
kc1kc4kc5=kc1*(1-lattice(jxi+k4))*(1-lattice(jx+k4))
nedges=nedges+kc1kc4kc5+kc2*(1-lattice(jx+Lx*(jyi+Ly*jzj)))*kc7 &
            +kc1*(1-lattice(jxi+k7))*kc7
if(kc1kc4kc5.ne.0) then
do k0=-1,1,2
jzk=jz+k0
k9=Lx*(jy+Ly*jzk)
k10=Lx*(jyj+Ly*jzk)
nvert=nvert+(1-lattice(jxi+k9))*(1-lattice(jxi+k10)) &
           *(1-lattice(jx+k10))*(1-lattice(jx+k9))
enddo ! k0
endif ! kc1kc4kc5
enddo ! j0
enddo ! i0

volume=volume_body
surface=surface_body+surface_face*nfaces
curvature=curv_body+curv_face*nfaces+curv_edge*nedges
euler3D=euler3D_body+euler3D_face*nfaces &
        +euler3D_edge*nedges+euler3D_vertex*nvert

end
```

Figure C.10.1 (***continued***).

Acknowledgments

The authors express their sincere gratitude to Kristel Michielsen and Hans DeRaedt for their inspirational collaborations and enthusiasm for foamy and porous media. Part of the content of the chapter is based upon the joint work published before, entitled *Aspects of mathematical topology* by Kristel F.L. Michielsen, Hans DeRaedt, Jeff Th.M. DeHosson, in *Advances in Imaging and Electron Physics* 125 119–195 (2002), Academic Press. Also, we are thankful to Freek Pasop of Skyscan for the micro-CT images and our team of students over the past years: Emiel Amsterdam, Sergey Punzhin, Oleksii Kuzmin, Diego Ribas Gomes, Leo de Jeer, Ronald Popma, and Paskal Kiers for their contributions to the study of structure—property relationships of porous material systems and for our mutual passionate discussions on the geometric and topological features of these phenomenal materials.

References

[1] P. Ball, The Self-Made Tapestry: Pattern Formation in Nature, Oxford University Press, Oxford, 1998.

[2] L. Euler, Elementa doctrinae solidorum. In: Novi Commentarii Academiae Scientiarum Petropolitanae, pp. 109–140, 1758.

[3] L.L. Bucciarelli, N. Dworsky, Sophie Germain, in: Sophie Germain. Studies in the History of Modern Science, vol. 6, Springer, Dordrecht, 1980, https://doi.org/10.1007/978-94-009-9051-7_2. Douglas P. Holmes Germain Curvature: The Case for Naming the Mean Curvature of a Surface after Sophie Germain, arXiv: 2303.13615v1 [math.HO] 23 Mar 2023.

[4] K.F.L. Michielsen, H. DeRaedt, J.Th.M. De Hosson, Aspects of mathematical morphology, Adv. Imag. Electron. Phys. 125 (2002) 119–195.

[5] J. Ohser, K. Schladitz, 3D Images of Materials Structures, Wiley-VCH Verlag GmbH, Weinheim, Germany, 2009.

[6] K.R. Castleman, Digital Image Processing, Prentice Hall, New Jersey, 1996.

[7] H. Hadwiger, Vorlesungen über Inhalt, Oberfläche und Isoperimetrie, Springer Verlag, Berlin, 1957.

[8] L.A. Santaló, Integral Geometry and Geometric Probability, Addison-Wesley, Reading, Massachusetts, 1976.

[9] D. Stoyan, W.S. Kendall, J. Mecke, Stochastic Geometry and its Applications, Akademie Verlag, Berlin, 1989.

[10] K.R. Mecke, H. Wagner, Euler characteristic and related measures for random geometric sets, J. Stat. Phys. 64 (1991) 843–850.

[11] K.R. Mecke, A morphological model for complex fluids, J. Phys. Condens. Matter 8 (1996) 9663–9667.

[12] K.R. Mecke, Morphological thermodynamics of composite media, Fluid Phase Equil. 150–151 (1998) 591–598.

[13] K.R. Mecke, Integral geometry in statistical physics, Int. J. Mod. Phys. B12 (1998) 861–899.

[14] A.L. Mellot, The topology of large-scale structure in the universe, Phys. Rep. 193 (1990) 1–39.

[15] K.R. Mecke, T. Buchert, H. Wagner, Robust morphological measures for large-scale structure in the Universe, Astron. Astrophys. 288 (1994) 697–704.

[16] N.G. Makarenko, L.M. Karimova, A.G. Terekhov, A.V. Kardashev, Minkowski functionals and comparison of discrete samples in seismology, Phys. Solid Earth 36 (2000) 305–309.

[17] J.S. Kole, K. Michielsen, H. De Raedt, Morphological image analysis of quan- tum motion in billiards, Phys. Rev. E63 (2001) 016201.

[18] C.N. Likos, K.R. Mecke, H. Wagner, Statistical morphology of random interfaces in microemulsions, J. Chem. Phys. 102 (1995) 9350–9361.

[19] K.R. Mecke, Morphological characterization of patterns in reaction-diffusionsystems, Syst. Phys. Rev. E53 (1996) 4794–4800.

[20] K.R. Mecke, V. Sofonea, Morphology of spinodal decomposition, Phys. Rev. E56 (1997) R3761–R3764.

[21] V. Sofonea, K.R. Mecke, Morphological characterization of spinodal decom- position kinetics, Eur. Phys. J. B8 (1999) 99–112.

[22] S. Herminghaus, K. Jacobs, K. Mecke, J. Bischof, A. Fery, M. Ibn-Elhaj, S. Schlagowski, Spinodal dewetting in liquid crystal and liquid metal films, Science 282 (1998) 916–919.

[23] K. Jacobs, S. Herminghaus, K.R. Mecke, Thin liquid polymer films rupture via defects, Langmuir 14 (1998) 965–969.

[24] G. Matheron, Random Sets and Integral Geometry, John Wiley and Sons, New York, 1975.
[25] J. Serra, Image Analysis and Mathematical Morphology, Academic Press, London, 1982.
[26] C.R. Giardina, E.R. Dougherty, Morphological Methods in Image and Signal Processing, Prentice Hall, New Jersey, 1988.
[27] I.R. Terol-Villalobos, Morphological image enhancement and segmentation, Adv. Imag. Electron. Phys. 118 (2001) 208–273.
[28] L.da F. Costa, R.M. Cesar Jr., Shape Analysis and Classification: Theory and Practice, CRC Press, Florida, 2001.
[29] K. Michielsen, H. De Raedt, Integral-geometry morphological image analysis, Phys. Rep. 347 (2001) 461–538.
[30] J. Klinowski, A.L. Mackay, H. Terrones, Curved surfaces in chemical structure, Phil. Trans. Roy. Soc. Lond. A354 (1996) 1975–1987.
[31] H.G. von Schnering, R. Nesper, How nature adapts chemical structures to curved surfaces, Angew. Chem. 26 (1987) 1059–1080.
[32] S. Andersson, S.T. Hyde, H.G. von Schnering, The intrinsic curvature of solids, Z. Kristallogr. 168 (1984) 1–17.
[33] A.L. Mackay, Periodic minimal surfaces, Nature 314 (1985) 604–606.
[34] A.L. Mackay, Equipotential surfaces in periodic charge distributions, Angew. Chem. 27 (1988) 849–850.
[35] I.S. Barnes, S.T. Hyde, B.W. Ninham, The caesium chloride zero potential surface is not the Schwarz P-surface, Coll. Phys. C7 (1990) 19–24.
[36] V. Luzzati, A. Tardieu, T. Gulik-Krzywicki, E. Rivas, F. Reiss-Husson, Structure of the cubic phases of lipid-water systems, Nature 220 (1968) 485–488.
[37] W. Longley, T.J. McIntosh, A bicontinuous tetrahedral structure in a liquid- crystalline lipid, Nature 304 (1983) 612–614.
[38] K. Larsson, Two cubic phases in monoolein-water system, Nature 304 (1983), 664–664.
[39] K. Fontell, Cubic phases in surfactant and surfactant-like lipid systems, Coll. Polym. Sci. 268 (1990) 264–285.
[40] V. Luzatti, R. Vargas, P. Mariani, A. Gulik, H. Delacroix, Cubic phases of lipid-containing systems. Elements of a theory and biological connotations, J. Mol. Biol. 229 (1993) 540–551.
[41] P. Mariani, V. Luzzati, H. Delacroix, Cubic phases of lipid-containing systems, J. Mol. Biol. 204 (1999) 165–189.
[42] E.L. Thomas, D.B. Alward, D.J. Kinning, D.C. Martin, D.L. Handlin Jr., L.J. Fetters, Ordered bicontinuous double-diamond structure of star block copolymers: a new equilibrium microdomain morphology, Macromolecules 19 (1986) 2197–2202.
[43] H. Hasegawa, H. Tanaka, K. Yamasaki, T. Hashimoto, Bicontinuous Microdomain morphology of block copolymers. 1. Tetrapod-network structure of polystyrene-polyisopropene diblock polymers, Macromolecules 20 (1987) 1651–1662.
[44] D.M. Anderson, E.L. Thomas, Microdomain morphology of star copolymers in the strong-segregation limit, Macromolecules 21 (1988) 3221–3230.
[45] E.L. Thomas, D.M. Anderson, C.S. Henkee, D. Hoffman, Periodic area-minimizing surfaces in block copolymers, Nature 334 (1988) 598–601.
[46] Y. Mogi, K. Mori, Y. Matsushita, I. Noda, Tricontinuous morphology of triblock copolymers of the ABC type, Macromolecules 25 (1992) 5412–5415.
[47] R. Xie, B. Yang, B. Jiang, The ordered bicontinuous double-diamond morphology in triblock copolymerhomopolymer blends, Macromolecules 26 (1993) 7097–7099.

[48] Y. Matsushita, M. Tamura, I. Noda, Tricontinuous double-diamond structure formed by a Styrene-Isopropene-2-Vinylpyridine Triblock copolymer, Macromolecules 27 (1994) 3680–3682.
[49] M.F. Schulz, F.S. Bates, K. Almdal, K. Mortensen, Epitaxial relationship for hexagonal-to-cubic phase transition in a block copolymer mixture, Phys. Rev. Lett. 73 (1994) 86–89.
[50] D.A. Hajduk, P.E. Harper, S.M. Gruner, C.C. Honeker, G. Kim, E.L. Thomas, L.J. Fetters, The gyroid: a new equilibrium morphology in weakly segregated diblock copolymers, Macromolecules 27 (1994) 4063–4075.
[51] D.A. Hajduk, P.E. Harper, S.M. Gruner, C.C. Honeker, E.L. Thomas, L.J. Fetters, A reevaluation of bicontinuous cubic phases in starblock copolymers, Macromolecules 28 (1995) 2570–2573.
[52] H. Jinnai, T. Hashimoto, D. Lee, S.-H. Chen, Morphological characterization of bicontinuous phase-separated polymer blends and one-phase microemulsions, Macromolecules 30 (1997) 130–136.
[53] S. Sakurai, H. Irie, H. Umeda, S. Nomura, H.H. Lee, J.K. Kim, Gyroid structures and morphological control in binary blends of polystyrene-*block*- polyisopropene diblock copolymers, Macromolecules 31 (1998) 336–343.
[54] M.W. Hamersky, M.A. Hillmeyer, M. Tirrell, F.S. Bates, T.P. Lodge, E.D. von Meerwall, Block copolymer self-diffusion in the gyroid and cylinder morphologies, Macromolecules 31 (1998) 5363–5370.
[55] M.E. Vigild, K. Almdal, K. Mortensen, I.W. Hamley, J.P.A. Fairclough, A.J. Ryan, Transformations to and from the gyroid phase in a diblock copolymer, Macromolecules 31 (1998) 5702–5716.
[56] F.S. Bates, G.H. Fredrickson, Block Copolymers-Designer Soft Materials. Advances in synthetic chemistry and statistical theory provide unparalleled control over molecular scale morphology in this class of macromolecules, Phys. Today 52 (1999) 32–38.
[57] H. Jinnai, T. Kajihara, H. Watashiba, Y. Nishikawa, R.J. Spontak, Interfacial and topological measurements of bicontinuous polymer morphologies, Phys. Rev. E64 (2001) 010803.
[58] G. Donnay, D.L. Pawson, X-Ray diffraction studies of echinoderm plates, Science 166 (1969) 1147–1150.
[59] H.-U. Nissen, Crystal orientation and plate structure in echinoid skeletal units, Science 166 (1969) 1150–1152.
[60] K. Larsson, Cubic lipid-water phases: structures and biomembrane aspects, J. Phys. Chem. 93 (1989) 7304–7314.
[61] R. Holyst, D. Plewczyński, A. Aksimentiev, K. Burdzy, Diffusion on curved, periodic surfaces, Phys. Rev. E60 (1999) 302–307.
[62] D. Plewczyński, R. Hol-yst, Reorientational angle distribution and diffusion coefficient for nodal and cylindrical surfaces, J. Chem. Phys. 113 (2000) 9920–9929.
[63] D. Plewczyński, R. Hol-yst, Approach to equilibrium of particles diffusing on curved surfaces, Physica A295 (2001) 371–378.
[64] W. Fischer, E. Koch, On 3-periodic minimal surfaces, Z. Kristallogr. 179 (1987) 31–52.
[65] K. Grosse-Brauckmann, On gyroid interfaces, J. Colloid Interface Sci. 187 (1997) 418–428.
[66] A.H. Schoen, Infinite Periodic Minimal Surfaces without Self-Intersections, NASA Technical NOTE No. D-5541, 1970.
[67] S.T. Hyde, The topology and geometry of infinite periodic surfaces, Z. Kristal- logr 187 (1989) 165–185.
[68] W. Fischer, E. Koch, Genera of minimal balance surfaces, Act. Cryst. A45 (1989) 726–732.

[69] A. Aksimentiev, M. Fial-kowski, R. Ho-lyst, Morphology of surfaces in mesoscopic polymers, surfactants, electrons, or reaction-diffusion systems: methods, simulations, and measurements, Adv. Chem. Phys. 121 (2002) 141–239.

[70] E. Detsi, E. De Jong, A. Zinchenko, Z. Vukovic, I. Vukovic, S. Punzhin, K. Loos, G. Ten Brinke, H.A. De Raedt, P.R. Onck, J.Th.M. De Hosson, On the specific surface area of nanoporous materials, Acta Mater. 59 (2011) 7488–7497.

[71] H.A. Schwarz, Gesammelte Mathematische Abhandlungen, Springer, Berlin, 1890.

[72] K. Michielsen, H. De Raedt, J.G.E.M. Fraaije, Morphological characterization of spatial patterns, Prog. Theor. Phys. Suppl. 138 (2000) 543–548.

[73] I. Vukovic, S. Punzhin, Z. Vukovic, P.R. Onck, J.Th.M. De Hosson, G. Ten Brinke, K. Loos, Supramolecular route to well-ordered metal nano-foams, Am. Chem. Soc. Nano 5 (2011) 6339–6346.

[74] E.R. Neovius, Bestimmung Zweier Spezieller Periodischer Minimalflächen, Akad. Abhandlungen, Helsingfors, 1883.

[75] E. Koch, W. Fischer, On 3-periodic minimal surfaces with non-cubic symmetry, Z. Krystallogr. 183 (1988) 129–152.

[76] D.J. Roaf, The Fermi surfaces of Cu, Ag, Au, Phil. Trans. Roy. Soc. Lond. A255 (1963) 135–152.

[77] A.L. Mackay, Crystallographic surfaces, Proc. Roy. Soc. Lond. A442 (1993) 47–59.

[78] H.G. von Schnering, R. Nesper, Nodal surfaces of Fourier series: fundamental invariants of structured matter, Z. Phys. B83 (1991) 407–412.

[79] U.S. Schwarz, G. Gompper, Systematic approach to bicontinuous cubic phases in ternary amphiphilic systems, Phys. Rev. E59 (1999) 5528–5541.

[80] P.J.F. Gandy, S. Bardhan, A.L. Mackay, J. Klinowski, Nodal surface approximations to the P, D, G and I-WP triply periodic minimal surfaces, Chem. Phys. Lett. 336 (2001) 187–195.

[81] J.C. Elliot, Method of Producing Metal Foam, vol 751, 1956, p. 289. US patent No. 2.

[82] E. Amsterdam, Structural Performance and Failure Analysis of Aluminium Foams, PhD Thesis, University of Groningen, April 18, 2008, p. 63, https://research.rug.nl/en/publications/structural-performance-and-failure-analysis-of-aluminium-foams.

[83] P.R. Onck, R. van Merkerk, A. Raaijmakers, J.Th.M. De Hosson, Fracture of open- and closed-cell metal foams, J. Materials Sci. 40 (2005) 5821–5828.

[84] V.S. Deshpande, M.F. Ashby, N.A. Fleck, Foam topology bending versus stretching dominated architectures, Acta Mater. 49 (2001) 1035–1040.

[85] E. Amsterdam, J.H.B. De Vries, J.Th.M. De Hosson, P.R. Onck, The influence of strain-induced damage on the mechanical response of open-cell aluminum foam, Acta Mater. 56 (2008) 609–618.

[86] E. Amsterdam, J.Th.M. De Hosson, P.R. Onck, Failure mechanisms of closed-cell aluminum foam under monotonic and cyclic loading, Acta Mater. 54 (2006) 4465–4472.

[87] A.G. Evans, J.W. Hutchinson, N.A. Fleck, M.F. Ashby, H.N.G. Wadley, The topological design of multifunctional cellular metals, Prog. Mater. Sci. 46 (2001) 309–327.

[88] Y. Nishikawa, H. Jinnai, T. Koga, T. Hashimoto, S.T. Hyde, Measurements of interfacial curvatures of bicontinuous structure from three-dimensional digital images. 1. A parallel surface method, Langmuir 14 (1998) 1242–1249.

[89] E.T. Faber, D. Martinez-Martinez, C. Mansilla, V. Ocelik, J. Th, M. De Hosson, Calibration-free quantitative surface topography reconstruction in scanning electron microscope, Ultramicroscopy 148 (2015) 31–41.

[90] E.T. Faber, Nanotopography and Morphology Along Metal-Polymer Interfaces, PhD Thesis, University of Groningen, December 19, 2014, p. 15, https://research.

rug.nl/en/publications/nanotopography-and-morphology-along-metal-polymer-interfaces.
[91] C. Mansilla, M.H. Novais, E. Faber, D. Martínez-Martínez, J.Th.M. De Hosson, On the 3D reconstruction of diatom frustules: a novel method, applications, and limitations, J. Appl. Phycol. 28 (2016) 1097–1110.
[92] R. Keller, R.H. Geiss, Transmission EBSD from 10 nm domains in a scanning electron microscope, J. Microsc. 245 (3) (2012) 245–251.
[93] R. van Bremen, D. Ribas Gomes, L.T.H. de Jeer, V. Ocelík, J.T.M. De Hosson, On the optimum resolution of transmission-electron backscattered diffraction (t-EBSD), Ultramicroscopy 160 (2016) 256–264.
[94] D. Ribas Gomes, Fabrication of Metallic Nanostructures With Ions, PhD Thesis, University of Groningen, October 9, 2018, p. 34, https://research.rug.nl/en/publications/fabrication-of-metallic-nanostructures-with-ions-theoretical-conc.
[95] L.T.H. de Jeer, D.R. Gomes, J.E. Nijholt, R. van Bremen, V. Ocelík, J.T.M. De Hosson, Formation of nanoporous gold studied by transmission electron backscatter diffraction, Microsc. Microanal. 21 (6) (2015) 1387–1397.
[96] L.T.H. de Jeer, PhD Thesis, Microscopy Study of Advanced Engineering Materials, 7, University of Groningen, January 12, 2018, p. 137, https://research.rug.nl/en/publications/microscopy-study-of-advanced-engineering-materials-crystallograph.
[97] H. Wang, H. Xiw, Y. Ju, Q. Duan, Error analysis of digital speckle correlation method under scanning electron microscope, Exp. Tech. 30 (2006) 42–45.
[98] M.A. Sutton, N. Li, D.C. Joy, A.P. Reynolds, X. Li, Scanning electron microscopy for quantitative small and large deformation measurements part I: SEM imaging at magnifications from 200 to 10,000, Exp. Mech. 47 (2007) 775–787.
[99] C. Mansilla, V. Ocelik, J.Th.M. De Hosson, A new methodology to analyze instabilities in SEM imaging, Microsc. Microanal. 20 (2014) 1625–1637.
[100] R.L.M. Popma, Sintering Characteristics of Nano-Ceramic Coatings, PhD Thesis, University of Groningen, April 29, 2002, p. 18, https://research.rug.nl/en/publications/sintering-characteristics-of-nano-ceramic-coatings.

Further reading

[1] M.F. Ashby, A.G. Evans, J.W. Hutchinson, N.A. Fleck, Metal Foams: A Design Guide, University of Cambridge, Cambridge UK, 1998.
[2] L. Gibson, M.F. Ashby, Cellular Solids, second ed., Cambridge University Press, Cambridge UK, 1997.
[3] A.E. Markaki, T.W. Clyne TW, The effect of cell wall microstructure on the deformation and fracture of aluminium-based foams, Acta Mater. 49 (2001) 1677–1686.
[4] E. Amsterdam, R. Goodall, A. Mortensen, P.R. Onck, J.Th.M. De Hosson, Fracture behavior of low-density replicated aluminum alloy foams, Mater. Sci. Eng. A 496 (2008) 376–382.
[5] E. Amsterdam, P.R. Onck, A. Raaijmakers, J.Th.M. De Hosson, Fracture and microstructure of open cell aluminum foam, J. Mater. Sci. 40 (2005) 5813–5819.

Index

'*Note:* Page numbers followed by "f" indicate figures and "t" indicate tables.'

9780443135378